SpringerBriefs in Computer Science

For further volumes:
http://www.springer.com/series/10028

Yindi Jing

Distributed Space-Time Coding

 Springer

Yindi Jing
University of Alberta
Edmonton, AB
Canada

ISSN 2191-5768   ISSN 2191-5776 (electronic)
ISBN 978-1-4614-6830-1   ISBN 978-1-4614-6831-8 (eBook)
DOI 10.1007/978-1-4614-6831-8
Springer New York Heidelberg Dordrecht London

Library of Congress Control Number: 2013931945

Printed on acid-free paper

Springer is part of Springer Science+Business Media (www.springer.com)

To my mother

Preface

Distributed space-time coding (DSTC) is a cooperative relaying scheme that enables high reliability in data transmission for wireless networks. It has been immensely successful and naturally attracted considerable attention from researchers since its introduction about one decade ago. Numerous publications have been devoted to this topic. However, an introductory text exclusively on this topic is still unavailable today, leaving aspiring researchers and students new to this field struggling with the scattered literature and sometimes confusing terminology. The goal of this book is to offer some help through accessible presentation of the basic ideas of DSTC as well as some related topics. The target audiences are researchers interested in getting to know DSTC, in particular graduate students. It is also my hope that this book can be useful to experts as a quick reference.

This book starts with an introduction on space-time coding for multiple-antenna system, cooperative relay network, and channel estimation. Then, the basic idea of DSTC and its performance analysis are presented, followed by generalizations, code design, and its differential use. Finally, recent results on channel estimation for DSTC and training-based DSTC are provided.

Some of the calculations and proofs involved are mathematical and can be safely skipped in the first reading. Nevertheless, I decided to include them because they either illustrate useful analytical skills or provide details that are missing in the original papers. Due to the limited time, space, and of course my knowledge and ability, the content of this book is far from extensive. It only includes closely related literatures that I am mostly familiar with.

I would like to express my greatest appreciation to Dr. Xuemin (Sherman) Shen for providing me the opportunity of writing this short book for Springer. I owe deep gratitude to my Ph.D. advisor Dr. Babak Hasibi and my post-doctoral supervisor Dr. Hamid Jafarkhani for introducing me to the field and guiding me in my research. I am grateful to all my collaborators and colleagues, in particular my former M.Sc. student Sun Sun who provided comments on Chap. 5. I also would like to thank Springer, especially Ms. Courtney Clark and Ms. Melissa Fearon, for their support in various aspects in the writing and publishing of this book. Finally,

I would like to acknowledge my husband, Dr. Xinwei Yu, who not only provided valuable comments on the writing of the book but also encouraged me throughout the process. The book would not have come into being without his support.

Edmonton, AB, Canada, December 2012   Yindi Jing

Contents

Acronyms

AF	Amplify-and-forward
BPSK	Binary phase-shift keying
CDF	Cumulative distribution function
CSI	Channel state information
DEC	Decoding
DF	Decode-and-forward
DPSK	Differential phase-shift keying
DSTC	Distributed space-time coding
gcd	Greatest common divider
i.i.d.	Independent and identically distributed
lcm	Least common multiplier
LMMSE	Linear minimum mean-square error
MIMO	Multiple-input-multiple-output
ML	Maximum-likelihood
MMSE	Minimum mean-square error
MRC	Maximum-ratio combining
MSE	Mean-square error
OD	Orthogonal design
PAM	Pulse-amplitude modulation
PEP	Pairwise error probability
PDF	Probability density function
PSK	Phase-shift keying
QAM	Quadrature amplitude modulation
QOD	Quasi-orthogonal design
QPSK	Quadrature phase-shift keying
Relay-RX	Relay-to-receiver
SISO	Single-input-single-output
SNR	Signal-to-noise ratio
SVD	Singular value decomposition
TDMA	Time division multiple access
TX-Relay	Transmitter-to-relay

Symbols

$\overline{(\cdot)}$	Complex conjugate
$(\cdot)^*$	Hermitian
$(\cdot)^t$	Transpose
$(\cdot)^\perp$	Orthogonal complement
$(\cdot)^{-1}$	Inverse
$\|\cdot\|_F$	Frobenius norm
$\|\cdot\|$	Absolute value
$\mathrm{tr}(\cdot)$	Trace
det	Determinant
rank	Rank
$\mathrm{vec}(\cdot)$	Vectorization of a matrix
$\Re(\cdot)$	Real part
$\Im(\cdot)$	Imaginary part
$\otimes$	Kronecker product
$\circ$	Hadamard product
$\mathbf{A} \preceq \mathbf{B}$	$\mathbf{B} - \mathbf{A}$ is a positive semi-definite matrix
$\subseteq$	Subset
$\mathbb{E}$	Expectation
$\mathbb{P}$	Probability
$\mathcal{CN}(m, \sigma^2)$	Circularly symmetric complex normal (Gaussian) distribution with mean m and variance σ^2
$\mathrm{diag}\{\mathbf{a}\}$	Diagonal matrix whose diagonal entries/blocks are elements of vector $\mathbf{a}$
$\mathrm{diag}\{a_1, \cdots, a_N\}$	Diagonal matrix whose diagonal entries/blocks are $a_1, \cdots, a_N$
$\log_e$	Natural logarithm
$\log_2$	Base-2 logarithm
$\log_{10}$	Base-10 logarithm
$\mathbf{0}$	Vector/matrix of all zeros
$\mathbf{I}$	Identity matrix
δ_{ij}	Kronecker delta. $\delta_{ij} = 1$ if $i = j$ and $\delta_{ij} = 0$ otherwise
$\hat{x}$	Estimation of x

P_s	Transmit power of the transmitter
P_r	Transmit power of each relay antenna
$\mathbf{n}_d, \mathbf{N}_d$	Noise vector, matrix at the destination
$\mathbf{n}_{r,i}$	Noise vector at the ith relay antenna
M	Number of antennas at the transmitter
N	Number of antennas at the receiver
R	Total number of antennas at the relays
$\mathbf{f}_i, f_i$	Channel vector, coefficient from the transmitter to the ith relay antenna
$\mathbf{f}$	Channel vector from the transmitter to all relay antennas
$\mathbf{g}_i, g_i$	Channel vector, coefficient from the ith relay antenna to the receiver
$\mathbf{G}$	Channel matrix from all relay antennas to the receiver
$\mathbf{B}, \mathbf{b}$	Information matrix, vector sent by the transmitter
$\mathbf{S}$	Distributed space-time codeword

Chapter 1
Introduction

Abstract This chapter provides a brief introduction to background materials. First, multiple-antenna system and space-time coding are reviewed in Sect. 1.1. Then we explain the motivation of cooperative relay network and a few widely used cooperative schemes in Sect. 1.2. In Sect. 1.3, we define and discuss in detail diversity order, a performance measure for reliability. A brief introduction on channel training in multiple-antenna system and relay network is provided in Sect. 1.4. In the last section of this chapter, notation used in this book are explained. References on space-time coding, cooperative relay network, and training in multiple-antenna system and cooperative relay network are provided at the end for the benefit of interested readers.

1.1 Multiple-Antenna System

Wireless communications have experienced several revolutions, including the appearance of AM and FM communication systems in early twentieth century and the development of the cellular phone systems from its first generation to the fourth generation in the last few decades. Next generation wireless communication systems are expected to meet considerably higher performance targets, for example, data rate at the order of gigabits per second, and reliability similar to that of wired networks. To meet these goals, traditional single-antenna systems, or single-input-single-output (SISO) system, is insufficient.

A significant technical breakthrough in the 1990s is the discovery of multiple-antenna system, or multiple-input-multiple-output (MIMO) system, which provides better capacity and reliability than SISO system. In a multiple-antenna system, more than one antenna is equipped at the transmitter and/or the receiver, resulting in multiple wireless channels between the transmitter and the receiver. This enables the communications between the transmitter and the receiver to benefit from multiple-path propagation, or fading, which is traditionally regarded as a disadvantage of a wireless channel. The fundamental limits of multiple-antenna system have been analyzed, including its capacity [7, 8, 12, 59, 66, 75], diversity gain

Y. Jing, *Distributed Space-Time Coding*, SpringerBriefs in Computer Science, DOI: 10.1007/978-1-4614-6831-8_1, © The Author(s) 2013

[1, 21–23, 44, 65, 69], and diversity-multiplexing tradeoff [50, 76]. Many transmission and reception techniques have been invented to efficiently achieve the high performance provided by the multiple transmit and receive antennas for different scenarios. Important ones include space-time coding [1, 21–23, 44, 63–65], beamforming [24, 41, 48, 51], and antenna selection [3, 15, 16, 20, 42, 47, 55, 56]. The focus of this book is the application of space-time coding idea in cooperative relay network. Thus the remaining part of this section is devoted to the background on space-time coding.

The idea of space-time coding was first proposed in [1, 65]. In space-time coding, information encoding is conducted not only in the time dimension, as is normally done in many single-antenna communication systems, but also in the spatial dimension. Redundancy is added coherently to both dimensions. By doing this, the data rate and the reliability can be largely improved with no extra cost on spectrum. This is also the main reason that space-time coding attracts significant attention from academic researchers and industrial engineers alike.

To understand the reliability improvement provided by space-time coding, we can look at the error probability. It has been proved in [65] that with space-time coding, the error probability of a multiple-antenna system with M antennas at the transmitter and N antennas at the receiver is proportional to SNR^{-MN}, where SNR is the signal-to-noise ratio (SNR). This is dramatic reduction compared with the error probability of a single-antenna system, which is proportional to SNR^{-1}. The negative of the exponent on the SNR, MN, is called the diversity order. It should be noted that this reliability improvement can be achieved only when the channels between the multiple transmit and receive antennas are rich-scattering (e.g., when the channels are independent) and follow Rayleigh fading. For correlated channels, the diversity order may be reduced.

The first practical space-time code was proposed by Alamouti in [1], which works for systems with two transmit antennas. It is also one of the most successful space-time codes because of its high performance but simple decoding. Later, a large variety of space-time codes have been designed.

The space-time coding scheme in [65] is based on the assumption that the receiver has full knowledge of the MIMO channels, which is not realistic for systems that suffer fast-changing channels. To solve this problem, differential space-time coding, a scheme that enables communications in systems with no channel information at either the transmitter or the receiver, was proposed in [22, 23]. Differential space-time coding can be seen as an extension of differential phase-shift keying (DPSK), a successful transmission scheme for single-antenna system that requires no channel state information (CSI) at the receiver, to multiple dimensions.

1.2 Cooperative Relay Network

In a multiple-antenna system, either the transmitter or the receiver or both are equipped with multiple antennas to achieve high performance. In many situations, however, due to limitation on size, processing power, and cost, it is not practical for some users, especially small wireless mobile devices, to be equipped with multiple antennas. The question of how to use MIMO techniques for single-antenna devices to achieve high performance thus arises.

On the other hand, a multiple-antenna system is a point-to-point communication system composed of two nodes only, one transmitter and one receiver. Communications between the transmitter and the receiver are direct, without relaying, cooperation, and/or interference from a third party. But for many modern applications and systems, communications are usually among multiple nodes, which form a wireless network, e.g., ad hoc wireless network and relay mesh network. The configurations of network communications also have grand potential in fulfilling the exploding demands of future communications.

These motivate the appearance of cooperative network and cooperative communications. The basic idea of cooperative communications is to have multiple single-antenna or multiple-antenna nodes in a network help on each other's communication tasks. With cooperative communications, antennas of different nodes can form a virtual MIMO system. By exploiting the spatial diversity provided by this virtual MIMO system, communication performance can be improved. It is an effective way to mitigate or even utilize the rich-scattering fading effect of wireless channels in a network.

One widely used model for cooperative network is cooperative relay network, where one node, called the transmitter, sends information to another node, called the receiver, while all other nodes in the network function as relays to help the transmitter-receiver pair. This network model is also called *single-user relay network*. If there are multiple nodes sending information with the help of some relay nodes, the network is called *multiple-user relay network*. This book is only concerned with single-user relay network.

One crucial question for cooperative communication is how the relays should help in the information transmission. A variety of cooperative relaying schemes have been proposed in the literature for different design criteria, applications, and network assumptions, including amplify-and-forward (AF) [2, 11, 13, 37, 49], decode-and-forward (DF) [2, 37, 57, 58], distributed space-time coding (DSTC) [26–29, 34, 52, 53], network beamforming [19, 30, 36, 39, 40, 43, 62, 70, 72], relay selection [5, 6, 31, 54, 73, 74], estimate-and-forward [17], coded-cooperation [25], etc.. We briefly explain some of these schemes in the following.

In AF, each relay or relay antenna simply amplifies its received signal from the transmitter based on its power constraint and forwards to the receiver. This scheme requires only simple processing at the relays. It however suffers the relay noise amplification as the relay noise is also directly forwarded to the receiver.

In DF, each relay or relay antenna conducts hard decoding of the information based on the signal it receives, then forwards its decoded information to the receiver either directly or after re-encoding. If correct decoding can be guaranteed at the relay, the DF scheme has high performance. But in practice, relays make decoding errors, so DF suffers from relay error propagation. Also, DF requires decoding at the relay, the complexity of which is usually high.

DSTC follows the idea of space-time coding proposed for multiple-antenna system. In this scheme, during a coherence interval, information is first sent from the transmitter to the relays as a multiple-dimensional vector. Each relay then linearly transforms its received signal vector with its distinctive transformation matrix, and forwards the processed signal vector to the receiver. At the receiver, a space-time codeword is formed without cross-talk among the relays or decoding at the relays. DSTC is shown to achieve the optimal diversity order with no CSI requirement at the transmitter or the relays.

Network beamforming uses the beamforming idea, where in amplifying the transmitter's information, each relay or relay antenna adjusts its transmit power and phase according to the channel status to form a directional beam toward the receiver. With beamforming, the optimal performance can be achieved. But it relies on precise and global CSI at the relays and the receiver, thus has heavy overhead and feedback load.

Relay selection lies in between DSTC and network beamforming. Its idea is to cleverly choose one or a subset of the available relays to help the communications. It requires partial/limited CSI (e.g., which relay provides the highest receive SNR), thus requires less overhead than beamforming but more overhead than DSTC. Its performance is naturally in-between network beamforming and DSTC.

1.3 Diversity Order and Pairwise Error Probability

There are many widely used performance measures for communication systems, for example, the capacity, outage probability, error probability, and energy efficiency. This book focuses on the error probability and analyzes its behaviour with respect to the average transmit power. The notation $\mathbb{P}(\text{error})$ is used to denote the probability of error, where $\mathbb{P}$ represents the probability. It can be the bit error rate or the symbol error rate. The notation P is used to denote the average transmit power. When the noise power is normalized as 1, it is also the average transmit SNR.

One quantitative representation for the error probability of a multiple-antenna system or a cooperative network is the *diversity order*, sometimes called *diversity gain* or simply *diversity*. In [76], it was defined as

$$d \triangleq - \lim_{P \to \infty} \frac{\log \mathbb{P}(\text{error})}{\log P}. \tag{1.1}$$

Diversity order describes the behaviour of the error probability with respect to the average transmit power in the log-log scale, when the power is asymptotically high. A scheme is said to achieve *full diversity* if the diversity order of its resulting error probability is the same as the maximum diversity available in the system.

For a multiple-antenna system, the error probability usually has the following behaviour

$$\mathbb{P}(\text{error}) \sim P^{-d},$$

where d is a constant independent of P. Here $f(x) \sim g(x)$ means $\lim_{x \to \infty} \frac{f(x)}{g(x)} = c$ with c a non-zero constant. In other words, $f(x)$ has the same scaling as $g(x)$ with respect to x when x is large. By using the definition in (1.1), the diversity order of the multiple-antenna system is d. For a cooperative relay network, however, the error probability may not always have such clean formula, for example, factors of $\log P$ may get involved. Thus, to have more precise representation of the error probability behaviour with respect to the average transmit power, we use the following definition of diversity order without the limit:

$$d = -\frac{\log \mathbb{P}(\text{error})}{\log P}. \tag{1.2}$$

Note that this definition may result in a diversity order that depends on the transmit power P, i.e., d may depend on P.

Diversity order shows how fast the error probability decreases as the transmit power increases. A larger diversity order means that the error probability can be reduced to a certain level with a smaller increase in the transmit power. Thus, it is important to guarantee the diversity order in a communication design.

The need for high diversity order in wireless communications can be seen as follows. For a wired communication link with the traditional additive white Gaussian noise channel model, the error probability behaves as $\mathbb{P}(\text{error}) \sim e^{-P}$, which decreases exponentially as the transmit power increases. The diversity order is thus infinity if the definition in (1.1) is used; and $P/\ln P$ if the definition in (1.2) is used. $P/\ln P$ is increasingly large for a large P. For example, when $P = 20\,\text{dB}$, the diversity order is 21.7. For a single wireless link modeled by Rayleigh fading, the error probability is shown to behaves as $\mathbb{P}(\text{error}) \sim P^{-1}$ [9], whose diversity order is 1. Thus, comparing the wired link with the wireless link, to achieve the same reliability improvement, a much higher transmit power is needed for the wireless link. This disadvantage can be overcome, to some extent, with diversity techniques. If by using multiple-antenna techniques, a diversity order of d (where $d > 1$) is achieved, the error probability will behave as P^{-d}. Compared with the single wireless link, the error rate decreases a lot faster as P increases.

If the error probability versus transmit power curve is drawn in the log-log scale, the diversity order can be read from the slope of the curve. In Fig. 1.1, the curves for a wired link (modeled as additive white Gaussian noise channel), a wireless link (modeled as Rayleigh flat-fading channel), and two systems whose diversity orders are 2 are shown. It can be seen that the slope of the curve for the wired link increases very fast, indicating an exponential decrease of the error probability with respect to the transmit power. For the other three, the curves mimic straight lines, showing a constant diversity order. The diversity order of the single wireless link is 1, and the diversity order of Design 1 and Design 2 are 2. Although with the same diversity order, Design 1 has a worse reliability than Design 2. This is possible when Design 2 has a better code design or takes advantages of clever combing scheme or selection scheme. Design 2 is usually said to have a better coding gain or array gain.

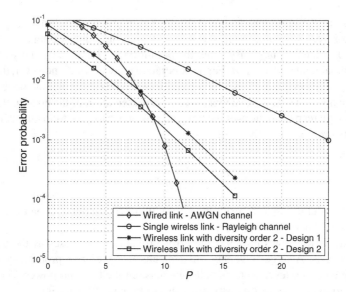

Fig. 1.1 Diversity order for a wired link, a wireless link, and two designs with diversity order 2

The calculation of the exact error probability is usually complicated. But, if the diversity order (not the exact value of the error probability) is the concern, one can resort to the calculation of the *pairwise error probability* (PEP). The PEP of mistaking one (scalar, vector, or matrix) codeword by another is the probability that the second codeword is decoded at the receiver while the first one is transmitted. It can be shown easily with the help of the union bound that for all practical digital modulation schemes, with respect to the average transmit power, the PEP and the exact error probability have the same scaling, thus the same diversity order. Since the calculation of the PEP is usually tidier, many investigations on diversity order work on the PEP instead.

1.4 Training and Channel Estimation

For many communication schemes, CSI is required at the receiver for information decoding. These schemes are usually called coherent schemes. For some designs, CSI is even required at the transmitter and relays (for relay network) for transmitter pre-processing and relay processing. To obtain such channel information, one common method is through a training process, where one or a sequence of pilot signals are sent by the transmitter. Based on the received signals, the receiver estimates the channels. Traditional schemes include maximum likelihood (ML) estimation, least-square estimation, minimum mean-square error (MMSE) estimation, and linear minimum mean-square error (LMMSE) estimation [33].

For a point-to-point SISO communication system, where there is one pair of transmitter and receiver, both equipped with one antenna, the training design and estimation scheme are straightforward. Traditional estimation schemes can be used directly and their performance has been analyzed [33]. For a multiple-antenna system, where there is one pair of transmitter and receiver, but both can be equipped with multiple antennas, the training design and estimation schemes have been well investigated in [4, 18, 67, 68]. The effect of channel estimation errors has also been studied in [38, 45, 71].

The training and channel estimation problems of cooperative relay network can be largely different to those of multiple-antenna system, due to the multiple-hop transmissions, distributed relays, and diverse relaying designs. Research on training design and channel estimation, along with the investigation on the effect of channel estimation errors on performance, are rapidly developing, e.g., [10, 14, 32, 35, 46, 60, 61]. In this book, several up-to-date training designs, channel estimation schemes, and training-based DSTC for cooperative relay network are introduced.

1.5 Explanation of Notation and Assumptions

The notation used in this book are as follows. Bold upper case letters are used to denote matrices and bold lower case letters are used to denote vectors, which can be either row vectors or column vectors. Letters with a regular font are scalars. For a matrix $\mathbf{A}$, its conjugate, transpose, Hermitian, orthogonal complement, Frobenius norm, determinant, rank, trace, and inverse are denoted by $\overline{\mathbf{A}}$, $\mathbf{A}^t$, $\mathbf{A}^*$, $\mathbf{A}^\perp$, $\|\mathbf{A}\|_F$, $\det(\mathbf{A})$, $\text{rank}(\mathbf{A})$, $\text{tr}(\mathbf{A})$, and $\mathbf{A}^{-1}$, respectively. $\text{vec}(\mathbf{A})$ denotes the column vector formed by stacking the columns of $\mathbf{A}$. $\mathbf{A} \succ 0$ means that $\mathbf{A}$ is positive definite. $\mathbf{A} \succ \mathbf{B}$ means that $\mathbf{A} - \mathbf{B}$ is positive definite. $\mathbf{A} \succeq \mathbf{B}$ means that $\mathbf{A} - \mathbf{B}$ is positive semi-definite. $\mathbf{I}_n$ is the $n \times n$ identity matrix. $\mathbf{0}$ is the all zero vector or matrix, whose dimension can usually be found out from the context. $\otimes$ denotes the Kronecker product. $\circ$ denotes the Hadamard product. For a complex scalar x, $|x|$ denotes its magnitude and $\angle x$ denotes its angle. $\Re(x)$ and $\Im(x)$ denote the real part and imaginary part of a complex scalar x. The same notation are used for the real and imaginary parts of a complex vector or matrix. $f(x) = \mathcal{O}(g(x))$ means $\lim_{x \to \infty} \frac{f(x)}{g(x)} = c$ with c a non-zero constant. $\mathbb{E}(\cdot)$ is the average operator for a random variable, a random vector, or a random matrix. $\mathbb{P}$ denotes the probability. $\hat{a}$ denotes the estimation of a. $a \triangleq b$ means that a is defined as b. $\mathcal{CN}(m, \sigma^2)$ is the circularly symmetric complex normal (Gaussian) distribution with mean m and variance σ^2. For a continuous random variable, $p_X(x)$ or $p(x)$ is its probability density function (PDF) and $P_X(x)$ or $P(x)$ is its cumulative distribution function (CDF). The same applies to a random vector.

Throughout this book, we assume that all channels experience Rayleigh flat-fading. With a *flat-fading* channel, all frequency components of the transmit signal experience the same magnitude of fading. This model is valid when the coherence bandwidth of the channel is larger than the bandwidth of the signal. All channels are assumed to be independent and identically distributed (i.i.d.) following Rayleigh distribution with parameter $\sigma^2 = 1/2$. Equivalently, all channel coefficients are independent and has circularly symmetric complex Gaussian distribution with zero-mean and unit-variance, denoted as $\mathcal{CN}(0, 1)$. Rayleigh fading model is justified for communications in an indoor or urban environment, where there is no line-of-sight. Channel independence can be guaranteed with enough spatial separation between antennas. For space-time coding or DSTC, we further use a *block-fading* model by assuming that the fading coefficients stay unchanged for T consecutive transmissions, then jump to independent values for another T transmissions and so on. T is called the *coherence interval*. Block-fading model has been widely used in wireless communications. It can approximately mimic the behaviour of a continuously fading process. It is also an accurate representation of many time division multiple access (TDMA), frequency-hopping, and block-interleaved systems. All noises are assumed to be i.i.d. circularly symmetric complex Gaussian, following $\mathcal{CN}(0, 1)$.

References

1. Alamouti SM (1998) A simple transmitter diversity scheme for wireless communications. IEEE J on Selected Areas in Communications, 16:1451–1458.
2. Azarian K, Gamal HE, and Schniter P (2005) On the achievable diversity-multiplexing tradeoff in half-duplex cooperative channels. IEEE T on Information Theory, 51:4152–4172.
3. Bahceci I, Duman TM, and Altunbasak Y (2003) Antenna selection for multiple-antenna transmission systems: performance analysis and code construction, IEEE T Information Theory, 49:2669–2681.
4. Biguesh M and Gershman AB (2006) Training-based MIMO channel estimation: A study of estimator tradeoffs and optimal training signals. IEEE T on Signal Processing, 54:884–893.
5. Bletsas A, Reed DP, and Lippman A (2006) A simple cooperative diversity method based on network path selection. IEEE J on Selected Areas in Communications, 24:659–672.
6. Bletsas A, Shin H, and Win MZ (2007) Outage optimality of opportunistic amplify-and-forward relaying. IEEE Communications L, 11:261–263.
7. Chizhik D, Foschini GJ, and Gans MJ (2002) Keyholes, correlations, and capacities of multi-element transmit and receive antennas. IEEE T on Wireless Communications, 1:361–368.
8. Chuah CN, Tse D, and Kahn JM (2002) Capacity scaling in MIMO wireless systems under correlated fading. IEEE T on Information Theory, 48:637–650.
9. Duman TM and Ghrayeb A (2007) *Coding for MIMO communication systems*. Wiley.
10. Gao F, Cui T, and Nallanathan A (2008) On channel estimation and optimal training design for amplify and forward relay networks. IEEE T Wireless Communications, 7:1907–1916.
11. F. Dana A and Hassibi B (2006) On the power-efficiency of sensory and ad-hoc wireless networks. IEEE T on Information Theory, 62:2890–2914.
12. Foschini GJ (1996) Layered space-time architecture for wireless communication in a fading environment when using multi-element antennas. Bell Labs Technical J, 1:41–59.
13. Gastpar M and Vetterli M (2002) On the capacity of wireless networks: the relay case. IEEE Infocom, 3:1577–1586.
14. Gedik B and Uysal M (2009) Impact of imperfect channel estimation on the performance of amplify-and-forward relaying. IEEE T on Wireless Communications, 8:1468–1479.
15. Gharavi-Alkhansari M and Greshman A (2004) Fast antenna selection in MIMO systems. IEEE T Signal Processing, 52:339–347.

16. Gorokhov A (2003) Receive antenna selection for MIMO spatial multiplexing: theory and algorithms, IEEE T Signal Processing, 51:2796–2807.
17. Gomadam KS and Jafar SA (2007) Optimal relay functionality for SNR maximization in memoryless relay networks. IEEE J on Selected Areas in Communications, 25:390–401.
18. Hassibi B and Hochwald BM (2003) How much training is needed in multiple-antenna wireless links? IEEE T on Information Theory, 49:951–963.
19. Havary-Nassab V, Shahbazpanahi S, Grami A, and Luo ZQ (2008) Distributed beamforming for relay networks based on second-order statistics of the channel state information. IEEE T Signal Processing, 56:4306–4316.
20. Heath RW Jr., Sandhu S, and Paulraj A (2001) Antenna selection for spatial multiplexing systems with linear receivers. IEEE Communications L, 5:142–144.
21. Hochwald BH and Marzetta TL (2000) Unitary space-time modulation for multiple-antenna communication in Rayleigh flat-fading. IEEE T on Information Theory, 46:543–564.
22. Hochwald BH and Sweldens W (2000) Differential unitary space-time modulation. IEEE T on Communications, 48:2041–2052.
23. Hughes B (2000) Differential space-time modulation. IEEE T on Information Theory, 46:2567–2578.
24. Jafar A and Goldsmith A (2004) Transmitter optimization and optimality of beamforming for multiple antenna systems. IEEE T Wireless Communications, 3:1165–1175.
25. Janani M, Hedayat A, Hunter TE, and Nosratinia A (2006) Coded cooperation in wireless communications: space-time transmission and iterative decoding. IEEE T on Signal Processing, 54:362–371.
26. Jing Y and Hassibi B (2006) Distributed space-time coding in wireless relay networks. IEEE T on Wireless Communications, 5:3524–3536.
27. Jing Y and Hassibi B (2008) Cooperative diversity in wireless relay networks with multiple-antenna nodes. EURASIP J on Advanced Signal Process, doi:10.1155/2008/254573.
28. Jing Y and Jafarkhani H (2007) Using orthogonal and quasi-orthogonal designs in wireless relay networks. IEEE T on Information Theory, 53:4106–4118.
29. Jing Y and Jafarkhani H (2008) Distributed differential space-time coding in wireless relay networks. IEEE T on Communications, 56:1092–1100.
30. Jing Y and Jafarkhani H (2009) Network beamforming using relays with perfect channel information. IEEE T on Information Theory, 55:2499–2517.
31. Jing Y and Jafarkhani H (2009) Single and multiple relay selection schemes and their diversity orders. IEEE T. on Wireless Communications, 8:1414–1423.
32. Jing Y and Yu X (2012) ML channel estimations for non-regenerative MIMO relay networks. IEEE J on Selected Areas in Communications, 30:1428–1439.
33. Kay SM (1993) *Fundamentals of Statistical Signal Processing, Volume I: Estimation Theory*, Prentice Hall.
34. Kiran T and Rajan BS (2007) Partial-coherent distributed space-time codes with differential encoder and decoder. IEEE J on Selected Areas in Communications, 25:426–433.
35. Kong T and Hua Y (2011) Optimal design of source and relay pilots for MIMO relay channel estimation. IEEE T on Signal Processing, 59:4438–4446.
36. Koyuncu E, Jing Y, and Jafarkhani H (2008) Distributed beamforming in wireless relay networks with quantized feedback, IEEE J on Selected Areas in Communications, 26:1429–1439.
37. Laneman JN and Wornell GW (2003) Distributed space-time-coded protocols for exploiting cooperative diversity in wireless network. IEEE T on Information Theory, 49:2415–2425.
38. Lapidoth A and Shamai A (2002) Fading channels: How perfect need perfect side information be? IEEE T Information Theory, 48:1118–1134.
39. Larsson P (2003), Large-scale cooperative relaying network with optimal combining under aggregate relay power constraint. Future Telecommunications Conference.
40. Li C and Wang X (2008) Cooperative multibeamforming in ad hoc networks. EURASIP J on Advances in Signal Processing, doi:10.1155/2008/310247.
41. Love DJ, Heath RW Jr, and Strohmer T (2003) Grassmannian beamforming for multiple-input multiple-output wireless systems. IEEE T on Information Theory, 49:2735–2747.

42. Mallik RK and Win MZ (2002) Analysis of hybrid selection/maximal-ratio combining in correlated Nakagami Fading. IEEE T Communications, 50:1372–1383.
43. Maric I and Yates RD (2010) Bandwidth and power allocation for cooperative strategies in Gaussian relay networks. IEEE T on Information Theory, 56:1880–1889.
44. Marzetta TL and Hochwald BH (1999) Capacity of a mobile multiple-antenna communication link in Rayleigh flat-fading. IEEE T on Information Theory, 45:139–157.
45. Medard M (2000) The effect upon channel capacity in wireless communications of perfect and imperfect knowledge of the channel. IEEE T Information Theory, 46:933–946.
46. Mheidat H and Uysal M (2007) Non-coherent and mismatched-coherent receivers for distributed STBCs with amplify-and-forward relaying. IEEE T on Wireless Communications, 6:4060–4070.
47. Molisch AF, Win MZ, Choi YS, and Winters JH (2005) Capacity of MIMO systems with antenna selection, IEEE T Wireless Communications, 4:1759–1772.
48. Mukkavilli KK, Sabharwal A, Erkip E, and Aazhang B (2003) On beamforming with finite rate feedback in multiple-antenna systems. IEEE T Information Theory, 49:2562–2579.
49. Nabar RU, Bolcskei H, and Kneubuhler FW (2004) Fading relay channels: Performance limits and space-time signal design. IEEE J on Selected Areas in Communications, 22:1099–1109.
50. Narasimhan R (2006) Finite-SNR diversity-multiplexing tradeoff for correlated Rayleigh and Rician MIMO channels. IEEE T on Information Theory, 52:3965–3979.
51. Narula A, Lopez MJ, Trott MD, and Wornell GW (1998) Efficient use of side information in multiple-antenna data transmission over fading channels. IEEE J on Selected Areas in Communications, 16:1423–1436.
52. Oggier F and Hassibi B (2010) Cyclic distributed space-time codes for wireless networks with no channel information. IEEE T on Information Theory, 56:250–265.
53. Rajan GS and Rajan BS (2007) Non-coherent low-decoding-complexity space-time codes for wireless relay networks, IEEE International Symposium on Information Theory, 1521–1525.
54. Ribeiro A, Cai X, and Giannakis GB (2005) Symbol error probabilities for general cooperative links. IEEE T on Wireless Communications, 4:1264–1273.
55. Sanayei S and Nosratinia A (2004) Antenna selection in MIMO systems. IEEE Communication M, 42:68–73.
56. Sanayei S and Nosratinia A (2007) Capacity of MIMO channels with antenna selection, IEEE T Information Theory, 53:4356–4362.
57. Sendonaris A, Erkip E, and Aazhang B (2003) User cooperation diversity-Part I: System description. IEEE T on Communications, 51:1927–1938.
58. Sendonaris A, Erkip E, and Aazhang B (2003) User cooperation diversity-Part II: Implementation aspects and performance analysis. IEEE T on Communications, 51:1939–1948.
59. Shiu DS, Foschini G, Gans M, and Kahn J (2000) Fading correlations and effect on the capacity of multielement antenna systems. IEEE T on Communications, 48:502–512.
60. Sun S and Jing Y (2011) Channel training design in amplify-and-forward MIMO relay networks. IEEE T on Wireless Communications, 10:3380–3391.
61. Sun S and Jing Y (2012) Training and decoding for cooperative network with multiple relays and receive antennas. IEEE T on Communications, 50:1534–1544.
62. Tang X and Hua Y (2007) Optimal design of non-regenerative MIMO wireless relays. IEEE T on Wireless Communications, 6:1398–1407.
63. Taricco G and Biglieri E (2005) Space-time decoding with imperfect channel estimation. IEEE T on Wireless Communications, 4:1874–1888.
64. Tarokh V, Naguib A, Seshadri N, and Calderbank AR (1999) Space-time codes for high-data-rate wireless communication: Performance criteria in the presence of channel estimation errors, mobility, and multiple paths. IEEE T on Communications, 47:199–207.
65. Tarokh V, Seshadri N, and Calderbank AR (1998) Space-time codes for high data rate wireless communication: performance criterion and code construction. IEEE T on Information Theory, 44:744–765.
66. Telatar IE (1999) Capacity of multi-antenna Gaussian channels. European T on Telecommunications, 10:585–595.

67. Tong L, Sadler BM, and Dong M (2004) Pilot-assisted wireless transmissions. IEEE T on Signal Processing, 21:12–25.

68. Vosoughi A and Scaglione A (2006) Everything you always wanted to know about training: Guidelines derived using the affine precoding framework and the CRB. IEEE T on Signal Processing, 54:940–954.

69. Wittneben A (1993) A new bandwidth efficient transmit antenna modulation diversity scheme for linear digital modulation. IEEE International Conference on, Communications, 1630–1634.

70. Yilmaz E and Sunay MO (2007) Amplify-and-forward capacity with transmit beamforming for MIMO multiple-relay channels. IEEE Globecom Conference, 3873–3877.

71. Yoo T and Goldsmith A (2006) Capacity and power allocation for fading MIMO channels with channel estimation error. IEEE T on Information Theory, 52:2203–2214.

72. Zhao Y, Adve R, and Lim TJ (2007) Beamforming with limited feedback in amplify-and-forward cooperative networks. IEEE Globecom Conference, 3457–3461.

73. Zhao Y, Adve R, and Lim TJ (2007) Improving amplify-and-forward relay networks: Optimal power allocation versus selection. IEEE T on Wireless Communications, 6:3114–3122.

74. Zhao Y, Adve R, and Lim TJ (2006) Symbol error rate of selection amplify-and-forward relay systems. IEEE Communications L, 10:757–759.

75. Zheng L and Tse D (2002) Communication on the Grassman manifold: a geometric approach to the noncoherent multiple-antenna channel. IEEE T on Information Theory, 48:359–383.

76. Zheng L and Tse D (2003) Diversity and multiplexing: A fundamental tradeoff in multiple-antenna channels. IEEE T on Information Theory, 49:1072–1096.

Chapter 2
Distributed Space-Time Coding

Abstract This chapter is on the distributed space-time coding (DSTC) scheme for cooperative relay network. At first, the space-time coding scheme proposed for multiple-antenna system is briefly reviewed in Sect. 2.1. Then in Sect. 2.2, DSTC for a single-antenna multiple-relay network is elaborated. The performance analysis of DSTC, including the pairwise error probability (PEP) calculations and the diversity order derivation, is also provided. Some code designs for DSTC are introduced in Sect. 2.3. Finally, in Sect. 2.4, simulated error probabilities of DSTC for several network scenarios are illustrated.

2.1 Review on Space-Time Coding

The idea of distributed space-time coding (DSTC) originated from space-time coding, a transmission scheme proposed for multiple-antenna system. In this section, we briefly review multiple-antenna system and space-time coding.

2.1.1 Multiple-Antenna System Model

A multiple-antenna system has two users. One is the transmitter and the other is the receiver. The transmitter has M transmit antennas and the receiver has N receive antennas, as illustrated in Fig. 2.1. There exists a wireless channel between each pair of transmit and receive antennas. With the Rayleigh flat-fading channel model, at the tth time slot, the channel between the mth transmit antenna and the nth receive antenna can be represented by a propagation coefficient, $h_{t,mn}$.

For each transmission time slot, one (uncoded or coded) signal can be sent from every transmit antenna. At the tth time slot, denote the signals to be sent by the M transmit antennas as $s_{t1}, \cdots, s_{tM}$, respectively. Let the average transmit power of the

Y. Jing, *Distributed Space-Time Coding*, SpringerBriefs in Computer Science,
DOI: 10.1007/978-1-4614-6831-8_2, © The Author(s) 2013

transmitter be P. The signals should be scaled by a power coefficient dependent in P before being sent. We denote it as $\sqrt{\beta_{\text{MIMO}}}$. Every antenna at the receiver receives a signal that is a superposition of the signals sent from all transmit antennas weighted by the corresponding channel coefficients. The received signal is also corrupted by noise. We denote the noise at the nth receive antenna by w_{tn}. The received signal at the nth receive antenna, denoted as x_{tn}, can be written as

$$x_{tn} = \sqrt{\beta_{\text{MIMO}}} \sum_{m=1}^{M} s_{tm} h_{t,mn} + w_n.$$

For the tth time slot, if we define the vector of the transmitted signals as $\mathbf{s}_t \triangleq [s_{t1}\ s_{t2}\ \cdots\ s_{tM}]$, the vector of the received signals as $\mathbf{x}_t \triangleq [x_{t1}\ x_{t2}\ \cdots\ x_{tN}]$, the vector of the noises as $\mathbf{w}_t \triangleq [w_{t1}\ w_{t2}\ \cdots\ w_{tN}]$, and the channel matrix as

$$\mathbf{H}_t \triangleq \begin{bmatrix} h_{t,11} & h_{t,12} & \cdots & h_{t,1N} \\ h_{t,21} & h_{t,22} & \cdots & h_{t,2N} \\ \vdots & \vdots & \ddots & \vdots \\ h_{t,M1} & h_{t,M2} & \cdots & h_{t,MN} \end{bmatrix},$$

the system equation of the multiple-antenna system can be written as

$$\mathbf{x}_t = \sqrt{\beta_{\text{MIMO}}} \mathbf{s}_t \mathbf{H}_t + \mathbf{w}_t. \tag{2.1}$$

2.1.2 Space-Time Coding

Space-time coding is a transmission scheme for multiple-antenna system to achieve the spatial diversity provided by the multiple antennas [1, 13, 30]. To use space-time coding, we assume the block-fading channel model with coherence interval T. When

Fig. 2.1 Multiple-antenna system

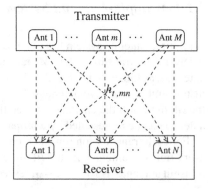

considering the transmissions within one coherence interval, the channel matrix $\mathbf{H}_t$ is the same for all t. Thus the index t can be omitted and the channel matrix can be denoted it as $\mathbf{H}$ only, i.e.,

$$
\mathbf{H} \triangleq
\begin{bmatrix}
h_{11} & h_{12} & \cdots & h_{1N} \\
h_{21} & h_{22} & \cdots & h_{2N} \\
\vdots & \vdots & \ddots & \vdots \\
h_{M1} & h_{M2} & \cdots & h_{MN}
\end{bmatrix},
$$

where h_{mn} is the channel coefficient between the mth transmit antenna and the nth receive antenna during the coherence interval.

Let

$$
\mathbf{S} \triangleq
\begin{bmatrix}
\mathbf{s}_1 \\
\mathbf{s}_2 \\
\vdots \\
\mathbf{s}_T
\end{bmatrix}
=
\begin{bmatrix}
s_{11} & s_{12} & \cdots & s_{1M} \\
s_{21} & s_{22} & \cdots & s_{2M} \\
\vdots & \vdots & \ddots & \vdots \\
s_{T1} & s_{T2} & \cdots & s_{TM}
\end{bmatrix}
$$

and

$$
\mathbf{X} =
\begin{bmatrix}
\mathbf{x}_1 \\
\mathbf{x}_2 \\
\vdots \\
\mathbf{x}_T
\end{bmatrix}
=
\begin{bmatrix}
x_{11} & x_{12} & \cdots & x_{1N} \\
x_{21} & x_{22} & \cdots & x_{2N} \\
\vdots & \vdots & \ddots & \vdots \\
x_{T1} & x_{T2} & \cdots & x_{TN}
\end{bmatrix},
\quad
\mathbf{W} =
\begin{bmatrix}
\mathbf{w}_1 \\
\mathbf{w}_2 \\
\vdots \\
\mathbf{w}_T
\end{bmatrix}
=
\begin{bmatrix}
w_{11} & w_{12} & \cdots & w_{1N} \\
w_{21} & w_{22} & \cdots & w_{2N} \\
\vdots & \vdots & \ddots & \vdots \\
w_{T1} & w_{T2} & \cdots & w_{TN}
\end{bmatrix}.
$$

For each coherence interval, by using the above matrix notation, we have from (2.1) the following system equation in matrix form:

$$
\mathbf{X} = \sqrt{\beta_{\mathrm{MIMO}}}\,\mathbf{SH} + \mathbf{W}. \tag{2.2}
$$

$\mathbf{S}$ is the transmitted signal matrix. Its (t, m)th entry, s_{tm}, is the signal sent by the mth transmit antenna at time t. The tth row of $\mathbf{S}$ is composed of signals sent by all transmit antennas at time t; and the mth column of $\mathbf{S}$ is composed of signals sent by the mth transmit antenna for all T time slots of the coherence interval. Therefore, the horizontal axis of $\mathbf{S}$ indicates the spatial domain and the vertical axis of $\mathbf{S}$ indicates the temporal domain. This is why $\mathbf{S}$ is called a space-time codeword, and the scheme is called *space-time coding*. In a space-time code design, information is coded coherently in both the spatial and the temporal domains.

In this book, we assume that the space-time codeword $\mathbf{S}$ is normalized as $\mathbb{E}\{\mathrm{tr}(\mathbf{S}^*\mathbf{S})\} = M$. Thus, the average total transmit power across all M antennas and all T time slots is

$$
\beta_{\mathrm{MIMO}}\mathbb{E}\{\mathrm{tr}(\mathbf{S}^*\mathbf{S})\} = \beta_{\mathrm{MIMO}}M.
$$

To have the average transmit power per time slot being P, we need

$$\beta_{\text{MIMO}} = \frac{PT}{M}.$$

$\mathbf{X}$ is the $T \times N$ matrix of the received signals. The tth row of $\mathbf{X}$ is composed of the received values at all receive antennas at time t and the nth column is composed of the received values of the nth receive antenna across the coherence interval. $\mathbf{W}$ is the $T \times N$ matrix of the noises.

Let S be the set of all space-time codewords/matrices. When the receiver knows the channel matrix $\mathbf{H}$, with independent and identically distributed (i.i.d.) Gaussian noises, the maximum-likelihood (ML) decoding of space-time coding is proved to be [9, 30]:

$$\arg \max_{S \in \mathcal{S}} \mathbb{P}(\mathbf{X}|\mathbf{S}) = \arg \min_{S \in \mathcal{S}} \left\| \mathbf{X} - \sqrt{\frac{PT}{M}} \mathbf{SH} \right\|_F^2. \tag{2.3}$$

The space-time coding scheme for multiple-antenna systems can be summarized as follows. First, information is encoded into $T \times M$ code-matrices. For each coherence interval, the transmitter selects a matrix in the codebook S according to the information bit string, and feeds columns of the matrix to its transmit antennas. The receiver receives a signal matrix, which is attenuated by channel fading and corrupted by noises, and decodes the information matrix. If the cardinality of S is L, the transmission rate of this code is $(\log_2 L)/T$ bits per transmission. The space-time code design problem is to design the set S.

2.1.3 Diversity Order of Space-Time Coding

In this subsection, we review results on the error probability and diversity order of space-time coding. The channels are assumed to be i.i.d. following circularly symmetric Gaussian distribution with zero-mean and unit-variance. Thus, the channel magnitude follow Rayleigh distribution.

For a multiple-antenna system with M transmit antennas and N receive antennas, with space-time coding, the pairwise error probability (PEP) of mistaking one space-time codeword $\mathbf{S}_k$ by another $\mathbf{S}_l$, averaged over the channel distribution, has the following upper bound [9, 11, 13, 30]:

$$\mathbb{P}(\mathbf{S}_k \rightarrow \mathbf{S}_l) \leq \det^{-N} \left[\mathbf{I}_M + \frac{PT}{4M} \mathbf{M}_{kl} \right], \tag{2.4}$$

where $\mathbf{M}_{kl} \triangleq (\mathbf{S}_k - \mathbf{S}_l)^*(\mathbf{S}_k - \mathbf{S}_l)$. It can be seen from the above formula that to minimize the PEP upper bound, $\mathbf{M}_{kl}$ should be full rank. That is, the set of code-matrices S should be design such that $\mathbf{S}_k - \mathbf{S}_l$ is full rank for all $\mathbf{S}_k, \mathbf{S}_l \in S$ and $\mathbf{S}_k \neq \mathbf{S}_l$. Such a code is said to be *fully diverse*.

If $T \geq M$ and the code is fully diverse, we have from (2.4),

$$\mathbb{P}(\mathbf{S}_k \to \mathbf{S}_l) \le \left(\frac{4M}{T}\right)^{MN} \det{}^{-N}(\mathbf{M}_{kl}) P^{-MN} + \mathcal{O}\left(P^{-MN-1}\right). \quad (2.5)$$

At high transmit power ($P \gg 1$), the first term in the right-hand-side of (2.5) is the dominant term. Therefore, from the definition in (1.2), diversity order MN is achieved by space-time coding. Since the multiple-antenna system has in total MN independent channel paths, the maximum spatial diversity order existing in the system is MN. This shows that space-time coding achieves full diversity. In general, with fully diverse code, the diversity order of space-time coding is $N \min\{T, M\}$.

The coefficient $\det(\mathbf{M}_{kl})$ in the PEP upper bound depends on the space-time code design. For the minimum error probability, this coefficient should be as large as possible. It is called the *coding gain*. For two code-designs with the same diversity order but different coding gains, their error probability curves have the same slope (parallel to each other) in the high power regime, but the code with a larger coding gain has lower error probability. More results on the error analysis of space-time coding can be found in [9, 13, 41, 42].

2.1.4 Differential Space-Time Coding

The previously discussed space-time coding scheme is a coherent scheme, where to conduct the ML decoding in (2.3), the receiver requires channel state information (CSI), that is, the receiver needs to know $\mathbf{H}$. No CSI is required at the transmitter. For some systems, however, neither the transmitter nor the receiver has CSI, due to fast-fading channels or the expensive cost in channel training and estimation. One method of communications in these systems is *differential space-time coding* [10, 11, 13], which can be seen as a higher-dimensional extension of differential phase-shift keying (PSK) commonly used in a signal-antenna system with no CSI.

In differential space-time coding, communications are in blocks of M transmissions, which implies that the transmitted signal is an $M \times M$ matrix. From Sect. 2.1.2, the transceiver equation at the τ-th block can be written as

$$\mathbf{X}_\tau = \sqrt{P}\mathbf{S}_\tau\mathbf{H}_\tau + \mathbf{W}_\tau, \quad (2.6)$$

where $\mathbf{S}_\tau$, $\mathbf{X}_\tau$, $\mathbf{H}_\tau$, and $\mathbf{W}_\tau$ are the transmitted signal matrix, the received signal matrix, the channel matrix, and the noise matrix at the τth block, respectively.

The transmitted matrix $\mathbf{S}_\tau$ is encoded differentially as the product of a unitary data matrix, $\mathbf{U}_{z_\tau}$ taken from the code-matrix set $\mathcal{U}$, and the transmitted matrix of the previous block $\mathbf{S}_{\tau-1}$. In other words,

$$\mathbf{S}_\tau = \mathbf{U}_{z_\tau}\mathbf{S}_{\tau-1}. \quad (2.7)$$

The transmitted matrix for the initial block can be set to be identity, i.e., $\mathbf{S}_0 = \mathbf{I}_M$. To assure the same transmit power for different blocks, $\mathbf{U}_{z_\tau}$ must be unitary. Since the channel is used in blocks of M time slots, the corresponding transmission rate is $(\log_2 L)/M$ bits per transmission, where L is the cardinality of $\mathcal{U}$.

If the channels keep constant for $2M$ consecutive channel uses, i.e., $\mathbf{H}_\tau = \mathbf{H}_{\tau-1}$, from the system equation in (2.6) and the differential encoding equation in (2.7), we have

$$\mathbf{X}_\tau = \sqrt{P}\mathbf{U}_{z_\tau}\mathbf{S}_{\tau-1}\mathbf{H}_{\tau-1} + \mathbf{W}_\tau = \mathbf{U}_{z_\tau}\mathbf{X}_\tau + \mathbf{W}_\tau - \mathbf{U}_{z_\tau}\mathbf{W}_{\tau-1}.$$

Therefore, the following differential system equation is obtained,

$$\mathbf{X}_\tau = \mathbf{U}_{z_\tau}\mathbf{X}_{\tau-1} + \mathbf{W}'_\tau, \tag{2.8}$$

where

$$\mathbf{W}'_\tau = \mathbf{W}_\tau - \mathbf{U}_{z_\tau}\mathbf{W}_{\tau-1}. \tag{2.9}$$

Since $\mathbf{U}_{z_\tau}$ is unitary, the noise term in (2.9), $\mathbf{W}'_\tau$, can be shown to be circularly symmetric Gaussian with distribution $\mathcal{CN}(\mathbf{0}, 2\mathbf{I}_M)$, or equivalently, entries of $\mathbf{W}'_\tau$ are i.i.d. $\mathcal{CN}(0, 2)$. Thus, $\mathbf{W}'_\tau$ is statistically independent of $\mathbf{U}_{z_\tau}$. Therefore, the ML decoding of differential space-time coding is

$$\arg\max_{\mathbf{U}\in\mathcal{U}} \|\mathbf{X}_\tau - \mathbf{U}\mathbf{X}_{\tau-1}\|_F^2. \tag{2.10}$$

No CSI is required for the receiver to conduct the decoding. It is shown in [10, 11] that in the high power regime, the average PEP of transmitting $\mathbf{U}_k$ and erroneously decoding $\mathbf{U}_l$ has the following upper bound

$$\mathbb{P}(\mathbf{U}_k \to \mathbf{U}_l) \le 8^{MN}|\det(\mathbf{U}_k - \mathbf{U}_l)|^{-2N}P^{-MN} + \mathcal{O}\left(P^{-MN-1}\right).$$

Thus, differential space-time coding also achieves full diversity order MN if the code-matrix set $\mathcal{U}$ is fully diverse. Its coding gain is $|\det(\mathbf{U}_k - \mathbf{U}_l)|^2$.

The system equation for differential space-time coding (2.8) has the same structure as that of coherent space-time coding in (2.2). But its equivalent noise term in (2.9) has twice the power. We can conclude that differential space-time coding has 3dB degradation in performance compared with coherent space-time coding.

2.2 Distributed Space-Time Coding for Single-Antenna Multiple-Relay Network

DSTC is basically the use of space-time coding scheme in cooperative relay network. It enables the distributed relay antennas function as antennas of the transmitter to form a virtual MIMO system. The DSTC scheme introduced in this book is based on the work in [14, 16–19].

2.2.1 Relay Network Model

We first introduce the relay network model. Consider a wireless relay network with one transmit node, one receive node, and R relays. A diagram of this network is shown in Fig. 2.2. The transmitter has single transmit antenna; and the receiver has single receive antenna. Each relay node has a single antenna that can be used for both transmission or reception. This network is also called a *single-antenna multiple-relay network*. The relay network whose nodes have multiple antennas is considered in the next chapter.

There is a wireless channel between the antenna of the transmitter and the antenna of each relay, and a wireless channel between the antenna of each relay and the antenna of the receiver. We assume that there is no direct channel between the transmitter and the receiver. The channels follow Rayleigh flat-fading model and block-fading model with coherence interval T.

By considering one coherence interval, the following notation for channels can be used. Denote the channel from the transmitter to the ith relay as f_i, and the channel from the ith relay to the receiver as g_i. f_i and g_i are assumed to be i.i.d. complex Gaussian with zero-mean and unit-variance, i.e., $\mathcal{CN}(0, 1)$. Thus the channel amplitude has Rayleigh distribution. Let

$$\mathbf{f} \triangleq \begin{bmatrix} f_1 \\ \vdots \\ f_R \end{bmatrix} \quad \text{and} \quad \mathbf{g} \triangleq \begin{bmatrix} g_1 \\ \vdots \\ g_R \end{bmatrix},$$

which are the channel vectors from the transmitter to the relays and from the relays to the receiver, respectively. In this chapter, we consider coherent transmission by assuming that the receiver has global CSI, i.e., the receiver knows all channels f_i's and g_i's. Its knowledge of the channels can be obtained by sending training signals from the relays and the transmitter, further details of which are provided in Chap. 5. The transmitter and relays have no CSI.

Fig. 2.2 Single-antenna multiple-relay network

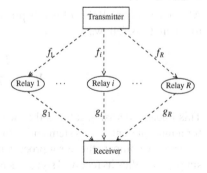

2.2.2 DSTC Transmission Protocol and Decoding

DSTC transmission takes two steps, where the first step is from the transmitter to the relays and the second step is from the relays to the receiver. The detailed formulations are as follows.

Step 1: Transmission from the Transmitter to the Relays

Step 1 is the transmission from the transmitter to the relays. It takes T time slots. Information bits are coded into a T-dimensional vector $\mathbf{b}$, normalized as

$$\mathbb{E}\{\mathbf{b}^*\mathbf{b}\} = 1. \tag{2.11}$$

Denote the ith entry of $\mathbf{b}$ as b_i. In the first step, the transmitter sends symbols $b_1, \cdots, b_T$ from time slot 1 to time slot T with the power coefficient $\sqrt{P_s T}$. By using the vector notation, this is equivalent to the transmitter sending $\sqrt{P_s T}\mathbf{b}$. Based on the normalization of $\mathbf{b}$ in (2.11), the total average transmit power of the T time slots in Step 1 can be calculated as

$$\mathbb{E}\left\{\left(\sqrt{P_s T}\mathbf{b}\right)^*\left(\sqrt{P_s T}\mathbf{b}\right)\right\} = P_s T.$$

Thus, for the transmitter, the average power per transmission is P_s. Denote the vector of the received signals at Relay i in this step as $\mathbf{r}_i$, which is also a T-dimensional column vector. Thus

$$\mathbf{r}_i = \sqrt{P_s T}\mathbf{b}f_i + \mathbf{n}_{r,i}, \tag{2.12}$$

where $\mathbf{n}_{r,i}$ is the T-dimensional noise vector at Relay i.

Relay Processing

After receiving $\mathbf{r}_i$, Relay i linearly processes $\mathbf{r}_i$ using a pre-determined $T \times T$ matrix to obtain $\mathbf{t}_i$ as follows:

$$\mathbf{t}_i = \sqrt{\frac{P_r}{P_s + 1}}\mathbf{A}_i\mathbf{r}_i. \tag{2.13}$$

This linear processing follows the idea of the linear dispersion space-time code [7] for a multiple-antenna system and is the essence of DSTC.

Now we compare the linear processing of DSTC with that of AF. In AF, each relay simply amplifies its received signals according to its power constraint. It corresponds to the case of DSTC where $\mathbf{A}_1 = \cdots = \mathbf{A}_R = \mathbf{I}_T$. With AF, the signal parts of $\mathbf{t}_i$'s are kept on the same subspace as the information vector **s**. For DSTC, the relays conduct more general linear transformations on the signals. Via proper $\mathbf{A}_i$ design, the information vector **s** can be mapped onto different subspaces by different relays, so the signal parts of $\mathbf{t}_i$'s are on different subspaces. The information is spread on both the temporal domain via the block transmission and the spatial domain via linear transformations at the relays. Information spreading in both the temporal and spatial domains is the key for space-time coding to achieve full diversity order.

Generally speaking, the $\mathbf{A}_i$ matrices can be arbitrary apart from a Frobenius norm constraint for the power normalization: $\text{tr}(\mathbf{A}_i^*\mathbf{A}_i) = T$. To have a protocol that is equitable among different relays and different time instants, we restrict $\mathbf{A}_i$'s to be unitary matrices. Since the relays have no channel knowledge and all channels are assumed to have identical distribution, an equitable design is optimal. Having unitary $\mathbf{A}_i$'s also simplifies the error probability analysis considerably without degrading the diversity order, which will be seen in later sections.

The coefficient $\sqrt{P_r/(P_s+1)}$ in (2.13) ensures that the power used at each relay for each transmission in the second step is P_r. The details will be explained in the succeeding description of the DSTC second step.

Step 2: Transmission from the Relays to the Receiver

The second transmission step of DSTC is from the relays to the receiver, where the ith relay sends the T-dimensional vector $\mathbf{t}_i$ to the receiver. This step also takes T time slots. All relays use the same spectral bandwidth. The relays are assumed to be synchronized, so their transmitted signals arrive at the receiver simultaneously. DSTC with asynchronous relays were investigated in [5, 6, 21].

Due to the normalization in (2.11) and the assumptions on the channels and the noises, we have

$$\mathbb{E}\{\mathbf{r}_i^*\mathbf{r}_i\} = \mathbb{E}\left\{\left(\sqrt{P_s T}\mathbf{b}f_i + \mathbf{n}_{r,i}\right)^*\left(\sqrt{P_s T}\mathbf{b}f_i + \mathbf{n}_{r,i}\right)\right\}$$
$$= \mathbb{E}\left\{P_s T\mathbf{b}^*\mathbf{b}|f_i|^2\right\} + \mathbb{E}\left\{\mathbf{n}_{r,i}^*\mathbf{n}_{r,i}\right\}$$
$$= P_s T\mathbb{E}\left\{|f_i|^2\right\}\mathbb{E}\left\{\mathbf{b}^*\mathbf{b}\right\} + T = (P_s + 1)T.$$

Therefore the average total transmit power at Relay i for all T transmissions in this step can be calculated to be

$$\mathbb{E}\{\mathbf{t}_i^*\mathbf{t}_i\} = \frac{P_r}{P_s+1}\mathbb{E}\{(\mathbf{A}_i\mathbf{r}_i)^*(\mathbf{A}_i\mathbf{r}_i)\} = \frac{P_r}{P_s+1}\mathbb{E}\{\mathbf{r}_i^*\mathbf{r}_i\} = P_r T.$$

P_r is thus the average transmit power per transmission for every relay during Step 2.

Denote the T-dimensional noise vector at the receiver as $\mathbf{n}_d$ and denote the T-dimensional received vector at the receiver as **x**. Thus,

$$\mathbf{x} = \sum_{i=1}^{R} \mathbf{t}_i g_i + \mathbf{n}_d = \begin{bmatrix} \mathbf{t}_1 & \cdots & \mathbf{t}_R \end{bmatrix} \mathbf{g} + \mathbf{n}_d. \tag{2.14}$$

All noises are assumed to be independent circularly symmetric complex Gaussian with zero-mean and unit-variance, that is, elements of $\mathbf{n}_{r,i}$, $\mathbf{n}_d$ follows i.i.d. $\mathcal{CN}(0, 1)$.

Transceiver Equation and Decoding of DSTC

By using the first step transceiver equation (2.12) and the relay processing design (2.13) in (2.14), we have

$$\mathbf{x} = \sqrt{\frac{P_r}{P_s+1}} \begin{bmatrix} \mathbf{A}_1 \mathbf{r}_1 & \cdots & \mathbf{A}_R \mathbf{r}_R \end{bmatrix} \mathbf{g} + \mathbf{n}_d$$

$$= \sqrt{\frac{P_s P_r T}{P_s+1}} \begin{bmatrix} \mathbf{A}_1 \mathbf{b} & \cdots & \mathbf{A}_R \mathbf{b} \end{bmatrix} \mathbf{f} \circ \mathbf{g}$$

$$+ \sqrt{\frac{P_r}{P_s+1}} \begin{bmatrix} \mathbf{A}_1 \mathbf{n}_{r,1} & \cdots & \mathbf{A}_R \mathbf{n}_{r,R} \end{bmatrix} \mathbf{g} + \mathbf{n}_d. \tag{2.15}$$

We use the following definitions:

$$\alpha \triangleq \frac{P_r}{P_s+1}, \qquad \beta \triangleq \frac{P_s P_r T}{P_s+1}, \tag{2.16}$$

$$\mathbf{S} \triangleq \begin{bmatrix} \mathbf{A}_1 \mathbf{b} & \cdots & \mathbf{A}_R \mathbf{b} \end{bmatrix}, \tag{2.17}$$

$$\mathbf{h} \triangleq \mathbf{f} \circ \mathbf{g} = \begin{bmatrix} f_1 g_1 \\ \vdots \\ f_R g_R \end{bmatrix}, \tag{2.18}$$

$$\mathbf{w} \triangleq \sqrt{\alpha} \begin{bmatrix} \mathbf{A}_1 \mathbf{n}_{r,1} & \cdots & \mathbf{A}_R \mathbf{n}_{r,R} \end{bmatrix} \mathbf{g} + \mathbf{n}_d. \tag{2.19}$$

The equation in (2.15) can be rewritten as

$$\mathbf{x} = \sqrt{\beta} \mathbf{S} \mathbf{h} + \mathbf{w}. \tag{2.20}$$

This system equation has the same structure as that of a multiple-antenna system in (2.2). By comparing the two, it can be seen that the $T \times R$ matrix $\mathbf{S}$ defined in (2.17) functions as the space-time code for the relay network. It is called the *distributed space-time code* since it is generated by the distributed relay antennas, without cross-talking and without decoding. Different columns of $\mathbf{S}$ are generated by different relay antennas. $\mathbf{h}$, which is $R \times 1$, is the end-to-end channel vector from the transmitter to the receiver via the relays. $\mathbf{w}$, which is $T \times 1$, is the equivalent noise. $\mathbf{w}$ is influenced by the relay linear transformation matrices $\mathbf{A}_1, \cdots, \mathbf{A}_R$, and the channels from the relays to the receiver, $g_1, \cdots, g_R$.

Since $\mathbf{A}_i$'s are unitary and entries of $\mathbf{n}_d$, $\mathbf{n}_{r,i}$ are i.i.d. standard complex Gaussian, for a given $\mathbf{g}$, $\mathbf{w}$ can be shown to be a circularly symmetric complex Gaussian vector. Its mean is $\mathbb{E}(\mathbf{w}) = \mathbf{0}_T$ and its covariance matrix is

$$
\begin{aligned}
\mathbf{R}_{\mathbf{w}} &\triangleq \mathbb{E}\{\mathbf{w}\mathbf{w}^*\} \\
&= \alpha\mathbb{E}\left\{\left[\mathbf{A}_1\mathbf{n}_{r,1} \quad \cdots \quad \mathbf{A}_R\mathbf{n}_{r,R}\right]\mathbf{g}\mathbf{g}^*\left[\mathbf{A}_1\mathbf{n}_{r,1} \quad \cdots \quad \mathbf{A}_R\mathbf{n}_{r,R}\right]^*\right\} + \mathbb{E}\{\mathbf{n}_d\mathbf{n}_d^*\} \\
&= \left(1 + \alpha\|\mathbf{g}\|_F^2\right)\mathbf{I}_T.
\end{aligned}
$$

This shows that $\mathbf{w}$ is both spatially and temporally white. Thus, for a given channel realization, when information vector $\mathbf{b}$ is sent (equivalently, when the space-time code matrix formed at the receiver is $\mathbf{S}$), the received vector $\mathbf{x}$ is Gaussian whose mean is $\sqrt{\beta}\mathbf{Sh}$ and whose variance is $\left(1 + \alpha\|\mathbf{g}\|_F^2\right)\mathbf{I}_T$. The conditional probability density function (PDF) of $\mathbf{x}|\mathbf{b}$ is

$$
p(\mathbf{x}|\mathbf{b}) = \left[\pi\left(1 + \alpha\|\mathbf{g}\|_F^2\right)\right]^{-T} e^{-\frac{\|\mathbf{x}-\sqrt{\beta}\mathbf{Sh}\|_F^2}{1+\alpha\|\mathbf{g}\|_F^2}}. \tag{2.21}
$$

The ML decoding of DSTC is thus

$$
\arg\max_{\mathbf{b}} p(\mathbf{x}|\mathbf{b}) = \arg\min_{\mathbf{b}} \left\|\mathbf{x} - \sqrt{\beta}\mathbf{Sh}\right\|_F^2. \tag{2.22}
$$

Recall that $\mathbf{S} = [\mathbf{A}_1\mathbf{b} \cdots \mathbf{A}_R\mathbf{b}]$, with $\mathbf{b}$ the information vector. By splitting the real and imaginary parts of the complex vectors in (2.22), the ML decoding is equivalent to

$$
\arg\min_{\mathbf{b}} \left\|\begin{bmatrix}\Re(\mathbf{x}) \\ \Im(\mathbf{x})\end{bmatrix} - \sqrt{\beta}\begin{bmatrix}\Re\left(\sum_{i=1}^R f_i g_i \mathbf{A}_i\right) & -\Im\left(\sum_{i=1}^R f_i g_i \mathbf{A}_i\right) \\ \Im\left(\sum_{i=1}^R f_i g_i \mathbf{A}_i\right) & \Re\left(\sum_{i=1}^R f_i g_i \mathbf{A}_i\right)\end{bmatrix}\begin{bmatrix}\Re(\mathbf{b}) \\ \Im(\mathbf{b})\end{bmatrix}\right\|_F^2. \tag{2.23}
$$

The ML decoding can be solved using sphere decoding [2, 8, 33], which has significantly reduced complexity compared with exhaustive search.

2.2.3 PEP Analysis of DSTC

In this section, the PEP of DSTC is calculated, from which the diversity order is derived.

We first calculate the PEP of mistaking one information vector $\mathbf{b}_k$ by another $\mathbf{b}_l$, denoted as $\mathbb{P}(\mathbf{b}_k \to \mathbf{b}_l)$. Let $\mathbf{S}_k$ and $\mathbf{S}_l$ be the space-time codewords corresponding to the information vectors $\mathbf{b}_k$ and $\mathbf{b}_l$, i.e.,

$$\mathbf{S}_k \triangleq \begin{bmatrix} \mathbf{A}_1 \mathbf{b}_k & \cdots & \mathbf{A}_R \mathbf{b}_k \end{bmatrix}, \quad \mathbf{S}_l \triangleq \begin{bmatrix} \mathbf{A}_1 \mathbf{b}_l & \cdots & \mathbf{A}_R \mathbf{b}_l \end{bmatrix}.$$

Define

$$\mathbf{M}_{kl} \triangleq (\mathbf{S}_k - \mathbf{S}_l)^*(\mathbf{S}_k - \mathbf{S}_l). \tag{2.24}$$

The following theorem has been proved.

Theorem 2.1. (Chernoff bound on the PEP). *[17] With the ML decoding in (2.22), the average PEP of mistaking $\mathbf{b}_k$ by $\mathbf{b}_l$ has the following upper bound:*

$$\mathbb{P}(\mathbf{b}_k \rightarrow \mathbf{b}_l) \leq \mathbb{E}_\mathbf{g} \det{}^{-1}\left[\mathbf{I}_R + \frac{\beta}{4\left(1 + \alpha \|\mathbf{g}\|_F^2\right)} \mathbf{M}_{kl} \mathrm{diag}\left\{ |g_1|^2, \cdots, |g_R|^2 \right\} \right]. \tag{2.25}$$

Proof.   With ML decoding, the PEP of mistaking $\mathbf{b}_k$ by $\mathbf{b}_l$ equals:

$$\mathbb{P}(\mathbf{b}_k \rightarrow \mathbf{b}_l) = \mathbb{P}(\log_e p(\mathbf{x}|\mathbf{b}_l) - \log_e p(\mathbf{x}|\mathbf{b}_k) > 0).$$

It has the following Chernoff upper bound [9, 32]:

$$\mathbb{P}(\mathbf{b}_k \rightarrow \mathbf{b}_l) \leq \mathbb{E}_{\mathbf{f},\mathbf{g},\mathbf{w}} e^{\lambda\left[\log_e p(\mathbf{x}|\mathbf{b}_l) - \log_e p(\mathbf{x}|\mathbf{b}_k)\right]}.$$

Since $\mathbf{b}_k$ is transmitted, $\mathbf{x} = \sqrt{\beta} \mathbf{S}_k \mathbf{h} + \mathbf{w}$. From (2.21),

$$\begin{aligned}
&\log_e p(\mathbf{x}|\mathbf{b}_l) - \log_e p(\mathbf{x}|\mathbf{b}_k) \\
&= -\frac{\|\mathbf{x} - \sqrt{\beta}\mathbf{S}_l\mathbf{h}\|_F^2 - \|\mathbf{x} - \sqrt{\beta}\mathbf{S}_k\mathbf{h}\|_F^2}{1 + \alpha \|\mathbf{g}\|_F^2} \\
&= -\frac{\|\sqrt{\beta}(\mathbf{S}_k - \mathbf{S}_l)\mathbf{h} + \mathbf{w}\|_F^2 - \|\mathbf{w}\|_F^2}{1 + \alpha \|\mathbf{g}\|_F^2} \\
&= -\frac{\beta \mathbf{h}^* \mathbf{M}_{kl} \mathbf{h} + \sqrt{\beta}\mathbf{h}^*(\mathbf{S}_k - \mathbf{S}_l)^* \mathbf{w} + \sqrt{\beta}\mathbf{w}^*(\mathbf{S}_k - \mathbf{S}_l)\mathbf{h}}{1 + \alpha \|\mathbf{g}\|_F^2}.
\end{aligned}$$

Therefore,

$$\begin{aligned}
&\mathbb{P}(\mathbf{b}_k \rightarrow \mathbf{b}_l) \\
&\leq \mathbb{E}_{\mathbf{f},\mathbf{g},\mathbf{w}} e^{-\frac{\lambda}{1 + \alpha \|\mathbf{g}\|_F^2}\left[\beta \mathbf{h}^* \mathbf{M}_{kl} \mathbf{h} + \sqrt{\beta}\mathbf{h}^*(\mathbf{S}_k - \mathbf{S}_l)^* \mathbf{w} + \sqrt{\beta}\mathbf{w}^*(\mathbf{S}_k - \mathbf{S}_l)\mathbf{h}\right]} \\
&= \mathbb{E}_{\mathbf{f},\mathbf{g}} \int \left[\pi \left(1 + \alpha \|\mathbf{g}\|_F^2\right) \right]^{-T} e^{-\frac{\lambda\left[\beta \mathbf{h}^* \mathbf{M}_{kl} \mathbf{h} + \sqrt{\beta}\mathbf{h}^*(\mathbf{S}_k - \mathbf{S}_l)^* \mathbf{w} + \sqrt{\beta}\mathbf{w}^*(\mathbf{S}_k - \mathbf{S}_l)\mathbf{h}\right] + \mathbf{w}\mathbf{w}^*}{1 + \alpha \|\mathbf{g}\|_F^2}} d\mathbf{w}
\end{aligned}$$

$$= \mathbb{E}_{\mathbf{f},\mathbf{g}} e^{-\frac{\lambda(1-\lambda)\beta}{1+\alpha\|\mathbf{g}\|_F^2}\mathbf{h}^*\mathbf{M}_{kl}\mathbf{h}} \int \left[\pi\left(1+\alpha\|\mathbf{g}\|_F^2\right)\right]^{-T} e^{-\frac{[\lambda\sqrt{\beta}(\mathbf{S}_k-\mathbf{S}_l)\mathbf{h}+\mathbf{w}]^*[\lambda\sqrt{\beta}(\mathbf{S}_k-\mathbf{S}_l)\mathbf{h}+\mathbf{w}]}{1+\alpha\|\mathbf{g}\|_F^2}} d\mathbf{w}$$

$$= \mathbb{E}_{\mathbf{f},\mathbf{g}} e^{-\frac{\lambda(1-\lambda)\beta}{1+\alpha\|\mathbf{g}\|_F^2}\mathbf{h}^*\mathbf{M}_{kl}\mathbf{h}}.$$

The last equality is because the integrand is a Gaussian PDF. Choose $\lambda = 1/2$, by which $\lambda(1-\lambda)$ reaches its maximum $1/4$. We have

$$\mathbb{P}(\mathbf{b}_k \to \mathbf{b}_l) \leq \mathbb{E}_{\mathbf{f},\mathbf{g}} e^{-\frac{\beta}{4(1+\alpha\|\mathbf{g}\|_F^2)}\mathbf{h}^*(\mathbf{S}_k-\mathbf{S}_l)^*(\mathbf{S}_k-\mathbf{S}_l)\mathbf{h}}. \tag{2.26}$$

To obtain (2.25), the expectation over $\mathbf{f}_i$ needs to be calculated. Notice that $\mathbf{h} = \text{diag}\{\mathbf{g}\}\mathbf{f}$. From (2.26),

$$\mathbb{P}(\mathbf{b}_k \to \mathbf{b}_l) \leq \mathbb{E}_{\mathbf{f},\mathbf{g}} e^{-\frac{\beta}{4(1+\alpha\|\mathbf{g}\|_F^2)}\mathbf{f}^*\text{diag}\{\mathbf{g}^*\}\mathbf{M}_{kl}\text{diag}\{\mathbf{g}\}\mathbf{f}}$$

$$= \mathbb{E}_{\mathbf{g}} \int \frac{1}{\pi^R} e^{-\frac{\beta}{4(1+\alpha\|\mathbf{g}\|_F^2)}\mathbf{f}^*\text{diag}\{\mathbf{g}^*\}\mathbf{M}_{kl}\text{diag}\{\mathbf{g}\}\mathbf{f}} e^{-\mathbf{f}^*\mathbf{f}} d\mathbf{f}$$

$$= \mathbb{E}_{\mathbf{g}} \det\left[\mathbf{I}_R + \frac{\beta}{4\left(1+\alpha\|\mathbf{g}\|_F^2\right)}\text{diag}\{\mathbf{g}^*\}\mathbf{M}_{kl}\text{diag}\{\mathbf{g}\}\right]^{-1}$$

$$= \mathbb{E}_{\mathbf{g}} \det\left[\mathbf{I}_R + \frac{\beta}{4\left(1+\alpha\|\mathbf{g}\|_F^2\right)}\mathbf{M}_{kl}\text{diag}\{[|g_1|^2 \cdots |g_R|^2]\}\right]^{-1}.$$

Power Allocation between the Transmitter and the Relays

Before further calculating the PEP, we look at the optimum power allocation between the transmitter and relays. For the tractability of analysis, we look for the power allocation that maximizes the average received SNR. First, since $\mathbf{A}_i$'s are unitary, from (2.11), we can derive the normalization of $\mathbf{S}$: $\text{tr}(\mathbf{S}^*\mathbf{S}) = R$. From the system Eq. (2.20), the average signal power in the received vector equals

$$\mathbb{E}\left\{\text{tr}\left[\beta\,(\mathbf{S}\mathbf{h})(\mathbf{S}\mathbf{h})^*\right]\right\} = \beta\mathbb{E}\{\text{tr}\left[\mathbf{S}\mathbb{E}(\mathbf{h}\mathbf{h}^*)\mathbf{S}^*\right]\} = \beta\mathbb{E}\{\text{tr}\left[\mathbf{S}\mathbf{S}^*\right]\} = R\beta.$$

The average noise power equals

$$\text{tr}(\mathbb{E}\{\mathbf{w}\mathbf{w}^*\}) = \text{tr}\left(\mathbb{E}\{1+\alpha\|\mathbf{g}\|_F^2\}\mathbf{I}_T\right) = T(1+R\alpha).$$

The received SNR is thus

$$\frac{R\beta}{T(1+R\alpha)} = \frac{R\frac{P_s P_r T}{P_s+1}}{T\left(1+R\frac{P_r}{P_s+1}\right)} = \frac{P_s P_r R}{P_s + RP_r + 1}.$$

Assume that the total power consumed in the whole network is P per symbol transmission. Thus $P = P_s + R P_R$, since for every symbol transmission, the powers used at the transmitter and every relay are P_s and P_r respectively. Therefore,

$$\frac{P_s P_r R}{P_s + P_r R + 1} = \frac{P_s(P - P_s)}{P + 1} \leq \frac{P^2}{4(P + 1)},$$

with equality when

$$P_s = \frac{P}{2} \quad \text{and} \quad P_r = \frac{P}{2R}. \tag{2.27}$$

Thus, the optimum power allocation is: the transmitter uses half the total power and the relays share the other half. For large R, the relays spend only a small amount of power to help the transmitter. A heuristic analysis on the power allocation that minimizes the PEP was provided in [17], which leads to the same result.

Chernoff Bound on the PEP

Now, we continue with the PEP calculation using the above optimal power allocation. The following theorem is derived.

Theorem 2.2. *Assume that $T \geq R$ and the distributed space-time code is fully diverse. With the power allocation in (2.27), the following bound on the PEP of DSTC can be derived:*

$$\mathbb{P}(\mathbf{b}_k \rightarrow \mathbf{b}_l) \leq \left(\frac{8R}{T}\right)^R \det{}^{-1}(\mathbf{M}_{kl}) \frac{\log_e^R P}{P^R} + \mathcal{O}\left(\frac{\log_e^{R-1} P}{P^R}\right). \tag{2.28}$$

Proof. Define $\gamma \triangleq R\left(1 + 2/P\right)$. With the power allocation in (2.27), from (2.25), we have

$$\mathbb{P}(\mathbf{b}_k \rightarrow \mathbf{b}_l) \leq \mathbb{E}_{\mathbf{g}} \det{}^{-1}\left[\mathbf{I}_R + \frac{PT}{8\left(\gamma + \|\mathbf{g}\|_F^2\right)}\mathbf{M}_{kl}\text{diag}\left\{\left[|g_1|^2 \cdots |g_R|^2\right]\right\}\right]. \tag{2.29}$$

Let $\lambda_i \triangleq |g_i|^2$. Therefore, λ_i is a random variable with exponential distribution. Its PDF is: $p(x) = e^{-x}$ for $x \geq 0$. From (2.29),

$$\mathbb{P}(\mathbf{b}_k \rightarrow \mathbf{b}_l) \leq \int_0^\infty \cdots \int_0^\infty \det{}^{-1}\left[\mathbf{I}_R + \frac{PT}{8\left(\gamma + \sum_{i=1}^R \lambda_i\right)}\mathbf{M}_{kl}\text{diag}\{[\lambda_1 \cdots \lambda_R]\}\right]$$
$$e^{-\lambda_1}\cdots e^{-\lambda_R}d\lambda_1 \cdots d\lambda_R. \tag{2.30}$$

Every integral in (2.30) can be broken into two parts: the integration from 0 to an arbitrary positive number x and the integration from x to ∞, to obtain

$$\mathbb{P}(\mathbf{b}_k \rightarrow \mathbf{b}_l) \leq \left(\int_0^x + \int_x^\infty \right) \cdots \left(\int_0^x + \int_x^\infty \right)$$

$$\det{}^{-1} \left[\mathbf{I}_R + \frac{PT}{8 \left(\gamma + \sum_{i=1}^R \lambda_i \right)} \mathbf{M}_{kl} \mathrm{diag}\{[\lambda_1 \cdots \lambda_R]\} \right] e^{-\lambda_1} \cdots e^{-\lambda_R} d\lambda_1 \cdots d\lambda_R$$

$$= \sum_{r=0}^R \sum_{1 \leq i_1 < \cdots < i_r \leq R} T_{i_1, \cdots, i_r}, \tag{2.31}$$

where

$$T_{i_1, \cdots, i_r} \triangleq \int \cdots \int_{\substack{\lambda_{i_1}, \cdots, \lambda_{i_r} \in (x, \infty) \\ \text{other } \lambda_i' s \in [0, x]}} \det{}^{-1} \left[\mathbf{I}_R + \frac{PT}{8 \left(\gamma + \sum_{i=1}^R \lambda_i \right)} \mathbf{M}_{kl} \mathrm{diag}\{[\lambda_1, \cdots, \lambda_R]\} \right]$$

$$e^{-\lambda_1} \cdots e^{-\lambda_R} d\lambda_1 \cdots d\lambda_R. \tag{2.32}$$

Without loss of generality, in what follows, $T_{1, \cdots, r}$ is calculated. Since the space-time code is fully diverse, $\mathbf{S}_k - \mathbf{S}_l$ is full rank, thus $\mathbf{M}_{kl} \succ \mathbf{0}$. Let $[\mathbf{M}_{kl}]_{1, \cdots, r}$ be the $r \times r$ matrix composed by the (i, j)th entries of $\mathbf{M}_{kl}$ for $i, j = 1, 2, \cdots, r$. When $x < \lambda_1, \cdots, \lambda_r < \infty$ and $0 < \lambda_{r+1}, \cdots, \lambda_R < x$,

$$\det \left[\mathbf{I}_R + \frac{PT}{8 \left(\gamma + \sum_{i=1}^R \lambda_i \right)} \mathbf{M}_{kl} \mathrm{diag}\{[\lambda_1, \cdots, \lambda_R]\} \right]$$

$$> \det \left[\mathbf{I}_R + \frac{PT}{8 \left[\gamma + (R - r)x + \sum_{i=1}^r \lambda_i \right]} \mathbf{M}_{kl} \mathrm{diag}\{[\lambda_1, \cdots, \lambda_r, 0, \cdots, 0]\} \right]$$

$$> \det \left[\frac{PT}{8 \left[\gamma + (R - r)x + \sum_{i=1}^r \lambda_i \right]} [\mathbf{M}_{kl}]_{1, \cdots, r} \mathrm{diag}\{[\lambda_1, \cdots, \lambda_r]\} \right]$$

$$= \left[\frac{PT}{8 \left[\gamma + (R - r)x + \sum_{i=1}^r \lambda_i \right]} \right]^r \lambda_1 \cdots \lambda_r \det[\mathbf{M}_{kl}]_{1, \cdots, r}.$$

Therefore,

$$
T_{1,\cdots,r} < \left(\frac{8}{PT}\right)^r \det{}^{-1}[\mathbf{M}_{kl}]_{1,\cdots,r} \left(\prod_{i=r+1}^{R} \int_0^x e^{-\lambda_i} d\lambda_i\right)
$$

$$
\int_x^\infty \cdots \int_x^\infty \left[\gamma + (R-k)x + \sum_{i=1}^{r} \lambda_i\right]^r \frac{e^{-\lambda_1}\cdots e^{-\lambda_r}}{\lambda_1\cdots\lambda_r} d\lambda_1\cdots d\lambda_r.
$$

(2.33)

From Lemma 1 in [17], it can be obtained that

$$
T_{1,\cdots,r} < \left(\frac{8}{PT}\right)^r \det{}^{-1}[\mathbf{M}_{kl}]_{1,\cdots,r} \left(1 - e^{-x}\right)^{R-r}
$$

$$
\sum_{j=0}^{r} B_{\gamma+(R-k)x,x}(j,r) \left[-Ei(-x)\right]^{r-j},
$$

(2.34)

where $\Gamma(\alpha, \chi)$ is the incomplete gamma function [4],

$$
B_{A,x}(j,k) \triangleq \binom{k}{j} \sum_{\substack{i_1,\ldots,i_j\geq 1 \\ \sum i_m \leq k}} C(i_1,\ldots,i_j)\Gamma(i_1, x)\cdots\Gamma(i_j, x)A^{r-i_1-\cdots-i_j},
$$

$$
C(i_1,\ldots,i_j) \triangleq \binom{k}{i_1}\binom{k-i_1}{i_2}\cdots\binom{k-i_1-\cdots-i_{j-1}}{i_j},
$$

and Ei is the exponential integral function [4].

In general, let $[\mathbf{M}_{kl}]_{i_1,\cdots,i_r}$ be the $r \times r$ matrix composed by the (i, j)th entries for $i, j = i_1, i_2, \cdots, i_r$ of $\mathbf{M}_{kl}$. Similarly, it can be proved that

$$
T_{i_1,\cdots,i_r} < \left(\frac{8}{PT}\right)^r \det{}^{-1}[\mathbf{M}_{kl}]_{i_1,\cdots,i_r} \left(1 - e^{-x}\right)^{R-r}
$$

$$
\sum_{j=0}^{r} B_{\gamma+(R-r)x,x}(j,r) \left[-Ei(-x)\right]^{r-j}.
$$

(2.35)

Thus, from (2.31), the PEP can be upper bounded as

$$
\mathbb{P}(\mathbf{b}_k \to \mathbf{b}_l) \leq \sum_{r=0}^{R} \left(\frac{8}{PT}\right)^r \left(\sum_{1 \leq i_1 < \cdots < i_r \leq R} \det{}^{-1}[\mathbf{M}_{kl}]_{i_1,\cdots,i_r}\right)
$$

$$
\left(1 - e^{-x}\right)^{R-r} \sum_{j=0}^{r} B_{\gamma+(R-r)x,x}(j,r) \left[-Ei(-x)\right]^{r-j}. \quad (2.36)
$$

Set $x = 1/P$. Notice that for large P,

$$
\begin{aligned}
&[\gamma + (R-k)/P]^k = \gamma^k + \mathcal{O}(1/P), \quad \gamma = R + \mathcal{O}(1/P) \\
&-Ei(-1/P) = \log_e P + \mathcal{O}(1) \qquad 1 - e^{-1/P} = 1/P + \mathcal{O}(1/P^2), \\
&\Gamma(i, 1/P) = (i-1)! + \mathcal{O}(1/P).
\end{aligned}
$$

The highest order term of P in the PEP upper bound in (2.36) is thus contained in the term where $r = R$ and $j = 0$, which is

$$
\begin{aligned}
&\left(\frac{8}{PT}\right)^R \det{}^{-1}(\mathbf{M}_{kl}) B_{\gamma,\frac{1}{P}}(0, r) \left[-Ei\left(-\frac{1}{P}\right)\right]^R \\
&= \left(\frac{8}{PT}\right)^R \det{}^{-1}(\mathbf{M}_{kl}) \gamma^R \left[\log_e P + \mathcal{O}(1)\right]^R \\
&= \left(\frac{8R}{T}\right)^R \det{}^{-1}(\mathbf{M}_{kl}) \left(\frac{\log_e P}{P}\right)^R + \mathcal{O}\left(\frac{\log_e^{R-1} P}{P^R}\right).
\end{aligned}
$$

The rest terms in the PEP upper bound in (2.36) has the scaling $\mathcal{O}\left(\frac{\log_e^{R-1} P}{P^R}\right)$ or lower. This proves the theorem.

Theorem 2.2 is very similar to Theorem 2 in [17]. But in Theorem 2.2, the PEP upper bound is represented to emphasize its highest order term with respect to P for high P, as diversity order is the major concern. In Theorem 2 of [17], the PEP upper bound is represented in further details, using Ei-function [4].

Diversity Gain of DSTC

When the transmit power is very high ($\log_e P \gg 1$),[1] the first term in the PEP formula in (2.28) is dominant. Since

$$
\frac{\log_e^R P}{P^R} = P^{-R\left(1 - \frac{\log_e \log_e P}{\log_e P}\right)},
$$

by using the definition in (1.2), an achievable diversity order of DSTC is

$$
d_{\text{DSTC}} = R\left(1 - \frac{\log_e \log_e P}{\log_e P}\right), \tag{2.37}
$$

which is linear in the number of relays. When the transmit power is large enough such that $\log_e P \gg \log_e \log_e P$, we have $\log_e \log_e P / \log_e P \ll 1$. Thus DSTC achieves diversity order R. Also, if the definition in (1.1) is adopted, an achievable diversity order of DSTC is R. Since there are in total R independent transmission paths in

---

[1] Note that, compared with the condition $P \gg 1$, this condition requires higher P.

the studied relay network, the maximum spatial diversity order of the network is R. Thus, we can conclude that DSTC achieves full diversity when the transmit power is very high.

In proving Theorem 2.2, it is assumed that $T \geq R$ and the distributed space-time code is fully diverse. In general, following a similar proof, it can be shown that an achievable diversity order of DSTC is

$$d_{\text{DSTC}} = \left(\min_{S_k, S_l \in \mathcal{S}} \text{rank}(\mathbf{M}_{kl}) \right) \left(1 - \frac{\log_e \log_e P}{\log_e P} \right).$$

If the code is fully diverse, we have

$$d_{\text{DSTC}} = \min\{T, R\} \left(1 - \frac{\log_e \log_e P}{\log_e P} \right).$$

In the proof of Theorem 2.2, (2.36) is shown to be a PEP upper bound of DSTC for any positive x. In obtaining the achievable diversity order in (2.37), x is set to be $1/P$. This actually is not the best choice of x according to diversity order. The optimal value of x was proved to be $P^{-\alpha_0}$ [15, 17], where α_0 satisfies

$$\alpha_0 + \frac{\log_e \alpha_0}{\log_e P} = 1 - \frac{\log_e \log_e P}{\log_e P}.$$

Bounds of α_0 were provided as

$$1 - \frac{\log_e \log_e P}{\log_e P} < \alpha_0 < 1 - \frac{\log_e \log_e P}{\log_e P} + \frac{\log_e \log_e P}{\log_e P(\log_e P - \log_e \log_e P)}.$$

With this choice of x, DSTC is shown to achieve diversity order $\alpha_0 R$. The detailed proof can be found in [15].

Coding Gain of DSTC

Theorem 2.2 also gives the coding gain of DSTC. When $\log_e P \gg 1$, the first term in the PEP formula in (2.28) is dominant. The coefficient $\det(\mathbf{M}_{kl})$ depends on the space-time code design. It is the coding gain of DSTC.

For a moderate P, the highest order term in the PEP formula is not dominant, thus the effect of the lower order terms cannot be neglected. From the proof of Theorem 2.2, it can be seen that the remaining terms in the PEP formula other than the highest order term have coefficients $\det([\mathbf{M}_{kl}]_{i_1,\cdots,i_r})$ for $\{i_1, \cdots, i_r\} \subseteq \{1, 2, \cdots, R\}$. Thus, for a low PEP, it is desirable to have $\det([\mathbf{M}_{kl}]_{i_1,\cdots,i_r})$ large for all $1 \leq r \leq R, 1 \leq i_1 < \cdots < i_r \leq R$. Notice that

$$[\mathbf{M}_{kl}]_{i_1,\cdots,i_r} = ([\mathbf{S}_k]_{i_1,\cdots,i_r} - [\mathbf{S}_l]_{i_1,\cdots,i_r})^*([\mathbf{S}_k]_{i_1,\cdots,i_r} - [\mathbf{S}_l]_{i_1,\cdots,i_r}),$$

where $[\mathbf{S}_k]_{i_1,\cdots,i_r} = [\mathbf{A}_{i_1}\mathbf{b}_k \cdots \mathbf{A}_{i_r}\mathbf{b}_k]$ and $[\mathbf{S}_l]_{i_1,\cdots,i_r} = [\mathbf{A}_{i_1}\mathbf{b}_l \cdots \mathbf{A}_{i_r}\mathbf{b}_l]$. Effectively, $[\mathbf{S}_k]_{i_1,\cdots,i_r}$ and $[\mathbf{S}_l]_{i_1,\cdots,i_r}$ are the space-time codewords corresponding to information vectors $\mathbf{b}_k$ and $\mathbf{b}_l$ when only the $i_1, \cdots, i_r$th relays are cooperating. So det $\left([\mathbf{M}_{kl}]_{i_1,\cdots,i_r}\right)$'s are large means that the distributed space-time code has large coding gain when any (arbitrary) subset of the relays cooperate and the rest do not cooperate. Thus, to have good performance when the transmit power is moderate, not only the distributed space-time code itself, but also any lower dimensional code formed by deleting some of the columns of the original code should have high coding gain.

Comparison with Multiple-Antenna System

Now, we compare DSTC in a single-antenna multiple-relay network with R relays with space-time coding in a multiple-antenna system with R transmit antennas and one receive antenna. The relay network model is shown in Fig. 2.2, while the multiple-antenna system is shown in Fig. 2.3.

First, both systems have R independent paths, thus the maximum spatial diversity order is R. For the multiple-antenna system, the diversity is provided by the R co-located transmit antennas; while for the relay network, the diversity is provided by the distributed relay antennas. To achieve spatial diversity, space-time coding is used for multiple-antenna system, where each co-located transmit antenna, knowing the transmit information perfectly, generate a column of the space-time codeword; while DSTC is used for the relay network, where each distributed relay antenna functions as a transmit antenna of the transmitter to generate a column of the distributed space-time codeword. For asymptotically high transmit power, both space-time coding and DSTC achieve full diversity order, which is R.

However, DSTC in relay network differs to space-time coding in multiple-antenna system in two aspects. One is that the noises at the relays are propagated to the receiver. The other is that every element in the equivalent channel vector is the product of two Rayleigh channel coefficients, due to the two-step transmission. These cause

Fig. 2.3 A multiple-antenna system with R transmit antennas and single receive antenna

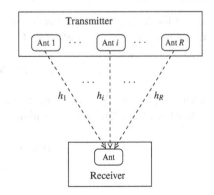

more difficulties in the performance analysis of DSTC, and lead to performance degradation. For finite transmit power, the diversity order of DSTC is lower than space-time coding by $(\log_e \log_e P)/\log_e P$. As the transmit power increases, the degradation vanishes.

Next, we compare the coding gain using the derived PEP Chernoff bounds of DSTC in (2.28) with that of space-time coding in (2.5) where $M = R, N = 1$. If only the dominant terms in the Chernoff bounds are considered, the two systems have exactly the same coding gain, which is $\det(\mathbf{M}_{kl})$. Thus, when the transmit power is high, linear dispersion space-time codes designed for multiple-antenna system can be employed to relay network directly. However, when the transmit power is not very high, for relay networks, a good distributed space-time code design should not only have high $\det(\mathbf{M}_{kl})$ but also have high $\det[\mathbf{M}_{kl}]_{i_1,\cdots,i_r}$ for all $1 \leq r \leq R, 1 \leq i_1 < \cdots < i_r \leq R$. The code design for DSTC will be studied in Sect. 2.3.

2.2.4 Generalized DSTC

In the previously introduced DSTC protocol, the two steps of transmissions have equal length. Also, in the relay processing, the transmit signal vector of each relay is a linear transformation of its received signal vector only. In this section, these requirements are relaxed and a more general structure of DSTC is introduced.

Consider a two-step transmission protocol where the first step takes T_1 time slots and the second step takes T_2 time slots. Assume that the coherence interval is no smaller than both T_1 and T_2, so that the channel coefficients keep unchanged during both transmission steps.

The same notation as in Sect. 2.2.2 are used in representing the protocol. Information is coded into a T_1-dimensional vector $\mathbf{b}$, normalized as $\mathbb{E}\{\mathbf{b}^*\mathbf{b}\} = 1$. In the first step, the transmitter sends $\sqrt{P_s T_1}\mathbf{b}$. Denote the T_1-dimensional vector Relay i receives as $\mathbf{r}_i$. This transmission step can be represented by the following equation:

$$\mathbf{r}_i = \sqrt{P_s T_1}\mathbf{b} f_i + \mathbf{n}_{r,i}. \tag{2.38}$$

For the relay processing, the ith relay linearly processes both its received signal $\mathbf{r}_i$ and the conjugate $\overline{\mathbf{r}_i}$ to generate $\mathbf{t}_i$ as:

$$\mathbf{t}_i = \sqrt{\alpha}(\mathbf{A}_i \mathbf{r}_i + \mathbf{B}_i \overline{\mathbf{r}_i}), \tag{2.39}$$

where $\mathbf{A}_i, \mathbf{B}_i$ are $T_2 \times T_1$ complex matrices and α is the relay power coefficient defined in (2.16). The transformation matrices are normalized as

$$\text{tr}(\mathbf{A}_i^* \mathbf{A}_i + \mathbf{B}_i^* \mathbf{B}_i) = T_2. \tag{2.40}$$

If we further assume that entries of $\mathbf{b}$ are zero-mean, uncorrelated, and satisfying $\mathbb{E}\{b_i^2\} = 0$, the relay transmit power per transmission can be shown to be P_r. Note that the conditions on $\mathbf{b}$ are not over-restrictive. For example, transmit vectors whose entries are i.i.d. phase-shift keying (PSK) signals satisfy the conditions. For a design of $\mathbf{b}$ that does not satisfy the condition, one can always adjust the normalization on $\mathbf{A}_i$ and $\mathbf{B}_i$ to have the average transmit power of each relay equal P_r.

After the relay processing in (2.39), in the second step, Relay i sends $\mathbf{t}_i$ to the receiver, simultaneously and using the same bandwidth. Notice that $\mathbf{t}_i$'s are T_2-dimensional vectors. The same notation $\mathbf{x}$ is used for the signal received at the receiver and $\mathbf{n}_d$ is used for the noise vector at the receiver. The transmission in Step 2 can be represented by (2.14).

Define

$$\beta_{\mathrm{gen}} \triangleq \frac{P_s P_r T_1}{P_s + 1}.$$

By using (2.38) and (2.39) in (2.14), similar to the derivation in Sect. 2.2.2, for this general DSTC design, the received signal vector can be written as,

$$\mathbf{x} = \sqrt{\beta_{\mathrm{gen}}} \left[\mathbf{A}_1 \mathbf{b} f_1 + \mathbf{B}_1 \overline{\mathbf{b} f_1} \quad \cdots \quad \mathbf{A}_R \mathbf{b} f_R + \mathbf{B}_R \overline{\mathbf{b} f_R} \right] \mathbf{g}$$
$$+ \sqrt{\alpha} \left[\mathbf{A}_1 \mathbf{n}_{r,1} + \mathbf{B}_1 \overline{\mathbf{n}_{r,1}} \quad \cdots \quad \mathbf{A}_R \mathbf{n}_{r,R} + \mathbf{B}_R \overline{\mathbf{n}_{r,R}} \right] \mathbf{g} + \mathbf{n}_d. \tag{2.41}$$

This equation has a similar structure to the transceiver equation (2.15) at the first sight. However, the matrix $\left[\mathbf{A}_1 \mathbf{b} f_1 + \mathbf{B}_1 \overline{\mathbf{b} f_1} \quad \cdots \quad \mathbf{A}_R \mathbf{b} f_R + \mathbf{B}_R \overline{\mathbf{b} f_R} \right]$ in the first term of the right-hand-side contains $f_1, \cdots, f_R$, channel coefficients from the transmitter to the relays, thus cannot be seen as the code-matrix. To separate the code from the channels, in what follows, an equivalent system equation is derived by separating the real and imaginary parts.

For any complex vector $\mathbf{x}$, define

$$\tilde{\mathbf{x}} \triangleq \begin{bmatrix} \Re(\mathbf{x}) \\ \Im(\mathbf{x}) \end{bmatrix}.$$

It thus follows from (2.39) that

$$\tilde{\mathbf{t}}_i = \sqrt{\alpha} \mathbf{C}_{i,\mathrm{gen}} \tilde{\mathbf{r}}_i, \tag{2.42}$$

where

$$\mathbf{C}_{i,\mathrm{gen}} \triangleq \begin{bmatrix} \Re(\mathbf{A}_i) + \Re(\mathbf{B}_i) & -\Im(\mathbf{A}_i) + \Im(\mathbf{B}_i) \\ \Im(\mathbf{A}_i) + \Im(\mathbf{B}_i) & \Re(\mathbf{A}_i) - \Re(\mathbf{B}_i) \end{bmatrix}, \tag{2.43}$$

which is a $(2T_2) \times (2T_1)$ matrix. We also introduce the following definitions:

$$\mathbf{f}_{\text{gen}} \triangleq \begin{bmatrix} \Re(f_1) & -\Im(f_1) & \cdots & \Re(f_R) & -\Im(f_R) \\ \Im(f_1) & \Re(f_1) & \cdots & \Im(f_R) & \Re(f_R) \end{bmatrix},$$

$$\mathbf{g}_{\text{gen}} \triangleq \begin{bmatrix} \Re(g_1) & -\Im(g_1) & \cdots & \Re(g_R) & -\Im(g_R) \\ \Im(g_1) & \Re(g_1) & \cdots & \Im(g_R) & \Re(g_R) \end{bmatrix},$$

$$\mathbf{n}_{\text{gen}} \triangleq \begin{bmatrix} \tilde{\mathbf{n}}_{r,1}^t & \cdots & \tilde{\mathbf{n}}_{r,R}^t \end{bmatrix}^t,$$

$$\mathbf{H}_{\text{gen}} = \left(\mathbf{g}_{\text{gen}} \otimes \mathbf{I}_T \right) \text{diag}\{\mathbf{C}_{1,\text{gen}}, \cdots, \mathbf{C}_{R,\text{gen}}\} \left(\mathbf{f}_{\text{gen}} \otimes \mathbf{I}_T \right),$$

$$\mathbf{w}_{\text{gen}} = \tilde{\mathbf{n}}_d + \sqrt{\alpha} \left(\mathbf{g}_{\text{gen}} \otimes \mathbf{I}_T \right) \text{diag}\{\mathbf{C}_{1,\text{gen}}, \cdots, \mathbf{C}_{R,\text{gen}}\}\mathbf{n}_{\text{gen}}.$$

By separating the real and imaginary parts of the system equation in (2.12) and (2.14), also using (2.42), after tedious but straightforward calculations, the following equivalent system equation can be obtained:

$$\tilde{\mathbf{x}} = \sqrt{\beta_{\text{gen}}}\mathbf{H}_{\text{gen}}\tilde{\mathbf{b}} + \mathbf{w}_{\text{gen}},$$

where $\mathbf{H}_{\text{gen}}$ and $\mathbf{w}_{\text{gen}}$ function as the equivalent channel matrix and equivalent noise vector, respectively. This transceiver equation can also be derived by separating the real and imaginary parts of the system equation in (2.41).

Assume that all noises are i.i.d. complex Gaussian with zero-mean and unit-variance; and the receiver has global CSI, i.e., the receiver knows both $\mathbf{f}, \mathbf{g}$. The ML decoding of the general DSTC can be shown straightforwardly to be

$$\arg\min_{\mathbf{b}} \left(\tilde{\mathbf{x}} - \sqrt{\beta_{\text{gen}}}\mathbf{H}_{\text{gen}}\tilde{\mathbf{b}} \right)^* \mathbf{R}_{\mathbf{w}_{\text{gen}}}^{-1} \left(\tilde{\mathbf{x}} - \sqrt{\beta_{\text{gen}}}\mathbf{H}_{\text{gen}}\tilde{\mathbf{b}} \right), \qquad (2.44)$$

where $\mathbf{R}_{\mathbf{w}_{\text{gen}}}$ is the covariance matrix of $\mathbf{w}_{\text{gen}}$. It can be derived from the definition of $\mathbf{w}_{\text{gen}}$ to be

$$\mathbf{R}_{\mathbf{w}_{\text{gen}}} \triangleq \frac{1}{2}\mathbf{I} + \frac{\alpha}{2} \left(\mathbf{g}_{\text{gen}} \otimes \mathbf{I}_{T_2} \right) \text{diag}\{\mathbf{C}_{1,\text{gen}}\mathbf{C}_{1,\text{gen}}^*, \cdots, \mathbf{C}_{R,\text{gen}}\mathbf{C}_{R,\text{gen}}^*\} \left(\mathbf{g}_{\text{gen}} \otimes \mathbf{I}_{T_2} \right)^*.$$

This ML decoding can be conducted using sphere decoding [3, 8, 34].

Diversity order analysis of the general DSTC has not been accomplished.

Other extensions of DSTC have been proposed in literature, e.g., [19, 24, 37]. For example, in [24], DSTC was extended to relay network with three transmission steps and in [19], DSTC designs when CSI is available at the relays were considered. In [37, 39, 40], the error probability of DSTC was analyzed.

A Special Case: Either $\mathbf{A}_i = 0$ or $\mathbf{B}_i = 0$

In what follows, a special case of the general DSTC is considered, which is widely used in DSTC research. Assume that $T_2 \geq T_1$ and for any i, either (1) $\mathbf{A}_i = \mathbf{0}$ and $\mathbf{B}_i$ is unitary or (2) $\mathbf{A}_i$ is unitary and $\mathbf{B}_i = \mathbf{0}$. That is, each relay linearly processes either its received vector or the conjugate of its received vector.

To simplify the representation, for this special case, define

$$\begin{cases} \breve{A}_i = A_i, \ \breve{f}_i = f_i, \ \breve{n}_{r,i} = n_{r,i}, \ b^{(i)} = b & \text{if } B_i = 0 \\ \breve{A}_i = B_i, \ \breve{f}_i = \overline{f}_i, \ \breve{n}_{r,i} = \overline{n}_{r,i}, \ b^{(i)} = \overline{b} & \text{if } A_i = 0 \end{cases}. \quad (2.45)$$

From (2.41), the system equation can again be written as (2.20), which is copied here for clear presentation

$$\mathbf{x} = \sqrt{\beta_{\text{gen}}} \mathbf{S} \mathbf{h} + \mathbf{w}, \quad (2.46)$$

where with slight abuse of notation, we re-define the distributed space-time code-word, the equivalent channel vector, and the equivalent noise vector as

$$\mathbf{S} \triangleq \begin{bmatrix} \breve{A}_1 b^{(1)} & \cdots & breve{A}_R b^{(R)} \end{bmatrix} \quad (2.47)$$

$$\mathbf{h} \triangleq \begin{bmatrix} \breve{f}_1 g_1 & \cdots & \breve{f}_R g_R \end{bmatrix}^t, \quad (2.48)$$

$$\mathbf{w} \triangleq \sqrt{\alpha} \sum_{i=1}^{R} g_i \breve{A}_i \breve{n}_{r,i} + \mathbf{n}_d, \quad (2.49)$$

respectively. It is straightforward to prove that $\mathbf{w}$ is a circularly symmetric complex Gaussian random vector whose mean is $\mathbf{0}$ and variance is

$$\mathbf{R_w} \triangleq \mathbf{I}_{T_2} + \alpha \sum_{i=1}^{R} |g_i|^2 \breve{A}_i \breve{A}_i^*. \quad (2.50)$$

The ML decoding of the system is therefore

$$\arg \min_{\mathbf{b}} \left(\mathbf{x} - \sqrt{\beta_{\text{gen}}} \mathbf{S} \mathbf{h} \right)^* \mathbf{R_w}^{-1} \left(\mathbf{x} - \sqrt{\beta_{\text{gen}}} \mathbf{S} \mathbf{h} \right). \quad (2.51)$$

An example of this special case is the use of Alamouti code [1] in a network with 2 relays, i.e., $R = 2$ and $T_1 = T_2 = 2$. If we design the relay processing matrices as $A_1 = I_2, B_1 = 0, A_2 = 0, B_2 = \begin{bmatrix} 0 & 1 \\ 1 & 0 \end{bmatrix}$, the distributed space-time codeword formed at the receiver has Alamouti structure [26].

Following similar arguments in Sect. 2.2.3, it can be shown that an achievable diversity order of this DSTC design is $\min\{T_2, R\} \left(1 - \frac{\log_e \log_e P}{\log_e P} \right)$ [26]. The optimal power allocation between the two transmission steps for the general DSTC was derived in [19].

2.3 Code Design for DSTC

In this section, we introduce several designs of distributed space-time code that lead to reliable communication in wireless relay network. The relay processing in (2.39), where either $\mathbf{A}_i = \mathbf{0}$, $\mathbf{B}_i$ is unitary or $\mathbf{B}_i = \mathbf{0}$, $\mathbf{A}_i$ is unitary, is considered. Note that this includes the basic DSTC in Sect. 2.2.2 by having $T_1 = T_2 = T$ and $\mathbf{B}_1 = \cdots = \mathbf{B}_R = \mathbf{0}$. Here, only the code designs in [19, 24] are presented. There are many other DSTC code designs in the literature, e.g., [20, 27, 38].

First, we review the code design criteria for DSTC. Let $\mathbf{b}_k$ and $\mathbf{b}_l$ be two information vectors. Their corresponding space-time codewords are thus,

$$\mathbf{S}_k \triangleq \left[\check{\mathbf{A}}_1 \mathbf{b}_k^{(1)} \cdots \check{\mathbf{A}}_R \mathbf{b}_k^{(R)} \right], \quad \mathbf{S}_l \triangleq \left[\check{\mathbf{A}}_1 \mathbf{b}_l^{(1)} \cdots \check{\mathbf{A}}_R \mathbf{b}_l^{(R)} \right].$$

It was shown in Sect. 2.2.3 that when the total transmit power in the whole network is very high (specifically $\log P \gg 1$), the coding gain of a distributed space-time code is $\det[(\mathbf{S}_k - \mathbf{S}_l)^*(\mathbf{S}_k - \mathbf{S}_l)]$. This is exactly the coding gain of the corresponding space-time code in a multiple-antenna system. It is thus natural to apply good space-time codes designs directly to relay networks.

However, for a general transmit power, it is shown in Sect. 2.2.3 that a good distributed space-time code should be "scale-free" in the sense that it still has a large coding gain when some columns of the code matrices are deleted, or equivalently, when some of the relays do not cooperate. This says that in addition to maximizing $\det[(\mathbf{S}_k - \mathbf{S}_l)^*(\mathbf{S}_k - \mathbf{S}_l)]$, one should also maximize

$$\det \left[([\mathbf{S}_k]_{i_1,\cdots,i_r} - [\mathbf{S}_l]_{i_1,\cdots,i_r})^* ([\mathbf{S}_k]_{i_1,\cdots,i_r} - [\mathbf{S}_l]_{i_1,\cdots,i_r}) \right], \tag{2.52}$$

for all $1 \leq r \leq R$, $1 \leq i_1 < \cdots < i_r \leq R$, where $[\mathbf{S}_k]_{i_1,\cdots,i_r} \triangleq \left[\check{\mathbf{A}}_{i_1} \mathbf{b}_k^{(i_1)} \cdots \check{\mathbf{A}}_{i_r} \mathbf{b}_k^{(i_r)} \right]$ is the sub-matrix of $\mathbf{S}_k$ composed of its $i_1, \cdots, i_r$th columns.

In the following, we present the orthogonal, quasi-orthogonal, and algebraic DSTC code designs.

2.3.1 Orthogonal Code

A (generalized) *complex OD* [13, 29] is a $T \times R$ matrix $\mathbf{S}$ whose entries are linear combinations of K complex indeterminate variables $b_1, \cdots, b_K$ and their conjugates, $b_1^*, \cdots, b_K^*$ such that

$$\mathbf{S}^*\mathbf{S} = \kappa \sum_{k=1}^{K} |b_k|^2 \mathbf{I}_R,$$

where κ is a constant. A (generalized) *real OD* [13, 29] is a $T \times R$ matrix $\mathbf{S}$ whose entries are linear combinations of K real indeterminate variables $b_1, \cdots, b_K$ such that

$$\mathbf{S}^t\mathbf{S} = \kappa \sum_{k=1}^{K} b_k^2 \mathbf{I}_R.$$

A set of code-matrices with the above unitary or orthogonal structure can be obtained if the information symbols $b_1, \ldots, b_K$ are selected from a modulation such as a PSK or a quadrature-amplitude modulation (QAM). This code-matrix set is called an orthogonal code. For real ODs, the modulation must be real. The symbol rate of the code is K/T.[2] If the cardinality of the constellation for each indeterminant variable is L, the total number of code-matrices is L^K. The bit rate of the DSTC transmission with this code will be $K \log_2 L/(2T)$ bits per time slot, as $2T$ time slots are needed for the transmission of one code-matrix.

ODs are ideal candidates for DSTC for the following reasons [19]. First, entries of codewords of an OD are linear in the information symbols and there conjugates. Therefore, they can be easily applied for DSTC. Second, ODs achieve full diversity and the largest coding gain [13, 29]. Third, due to the special structure, ODs may have simple symbol-wise decoding, where in the ML decoding formula, information symbols can be decoupled completely [13, 29] so the decoding of one symbol is independent on that of another. Finally, ODs have the "scale-free" property. If some columns of the code-matrices are deleted, the remaining lower-dimensional code is still an orthogonal code with full diversity order and the largest coding gain. This can be straightforwardly seen from the definitions of ODs. If $\mathbf{S}_k$ and $\mathbf{S}_l$ are two elements of an real or complex orthogonal code, we have

$$\det([\mathbf{S}_k]_{i_1,\cdots,i_r} - [\mathbf{S}_l]_{i_1,\cdots,i_r})^*([\mathbf{S}_k]_{i_1,\cdots,i_r} - [\mathbf{S}_l]_{i_1,\cdots,i_r}) = \kappa^r \left(\sum_{k=1}^{T} |b_k|^2 \right)^r$$

for any $1 \leq r \leq R$ and $1 \leq i_1 < \cdots < i_r \leq R$.

Application of Real ODs

For a real OD, every entry of the code-matrix is a linear combination of the information symbols. The ith column of the code-matrix can be written as $\mathbf{A}_i\mathbf{b}$ naturally. Actually, from the definition of real ODs, it is easy to prove that $\mathbf{A}_i$ satisfies

---

[2] For a multiple-antenna system that uses this code in space-time coding, K symbols can be sent through T time slots. Thus the symbol-rate is K/T. For a relay network with DSTC, $2T$ time slots are actually needed to complete the transmissions of the K symbols. But to be consistent with the literature on ODs, we say that the code has symbol-rate K/T.

$$\begin{cases} \mathbf{A}_i^t \mathbf{A}_i = \mathbf{I}_T \\ \mathbf{A}_i^t \mathbf{A}_j = -\mathbf{A}_j^t \mathbf{A}_i \end{cases} \cdot \qquad (2.53)$$

To apply a real OD to DSTC, we design the vector sent by the transmitter to be $\mathbf{b} = [b_1 \cdots b_K]^t$ and the processing matrix of the ith relay to be $\mathbf{A}_i$, while setting $\mathbf{B}_i = \mathbf{0}$. The first equation in (2.53) ensures $\mathbf{A}_i$ being unitary.

As an example, in what follows, we explain the applications of the 2×2 and the 4×4 real ODs [19]:

$$\mathbf{S} = \begin{bmatrix} b_1 & b_2 \\ -b_2 & b_1 \end{bmatrix} \quad \text{and} \quad \mathbf{S} = \begin{bmatrix} b_1 & -b_2 & b_3 & b_4 \\ b_2 & b_1 & -b_4 & b_3 \\ b_3 & -b_4 & -b_1 & -b_2 \\ b_4 & b_3 & b_2 & -b_1 \end{bmatrix}.$$

The 2×2 OD can be used in networks with 2 relays, i.e., $R = 2$. Let $T_1 = T_2 = 2$ and $K = 2$. Let b_1 and b_2 be the two information symbols to be sent, which can be modulated using real constellations such as pulse-amplitude modulation (PAM) with proper normalization. Design the 2×1 vector to be transmitted by the transmitter and the 2 relay processing matrices as

$$\mathbf{b} = \begin{bmatrix} b_1 \\ b_2 \end{bmatrix}, \mathbf{A}_1 = \begin{bmatrix} 1 & 0 \\ 0 & -1 \end{bmatrix}, \mathbf{A}_2 = \begin{bmatrix} 0 & 1 \\ 1 & 0 \end{bmatrix}, \mathbf{B}_1 = \mathbf{B}_2 = \mathbf{0}.$$

From (2.17), it can be seen that the 2×2 space-time codeword formed at the receiver has the desired real OD structure.

The 4×4 real OD can be used in networks with 4 relays, i.e., $R = 4$. Let $T_1 = T_2 = 4$, and $K = 4$. Let b_1, b_2, b_3, b_4 be the four information symbols to be sent. Design the 4×1 vector to be sent by the transmitter and the relay processing matrices as

$$\mathbf{b} = \begin{bmatrix} b_1 \\ b_2 \\ b_3 \\ b_4 \end{bmatrix}, \quad \mathbf{A}_1 = \mathbf{I}_4, \quad \mathbf{A}_2 = \begin{bmatrix} 0 & -1 & 0 & 0 \\ 1 & 0 & 0 & 0 \\ 0 & 0 & 0 & -1 \\ 0 & 0 & 1 & 0 \end{bmatrix}$$

$$\mathbf{A}_3 = \begin{bmatrix} 0 & 0 & 1 & 0 \\ 0 & 0 & 0 & -1 \\ -1 & 0 & 0 & 0 \\ 0 & 1 & 0 & 0 \end{bmatrix}, \quad \mathbf{A}_4 = \begin{bmatrix} 0 & 0 & 0 & 1 \\ 0 & 0 & 1 & 0 \\ 0 & -1 & 0 & 0 \\ -1 & 0 & 0 & 0 \end{bmatrix},$$

$$\mathbf{B}_1 = \mathbf{B}_2 = \mathbf{B}_3 = \mathbf{B}_4 = \mathbf{0}.$$

From (2.17), it can be seen that the 4×4 space-time codeword formed at the receiver has the desired real OD structure. Other real ODs can be applied similarly.

Application of Complex ODs

In a complex OD, not only the information symbols but also their conjugates appear in the code-matrix. For the special DSTC where either $\mathbf{A}_i = \mathbf{0}$, $\mathbf{B}_i$ is unitary or $\mathbf{B}_i = \mathbf{0}$, $\mathbf{A}_i$ is unitary (equivalently, a relay linearly transforms either its received vector or the conjugate of its received vector, but not both), each column of the code-matrix can contain linear transforms of either the information symbols, $b_1, \cdots, b_K$, or their conjugates, $b_1^*, \ldots, b_K^*$, but not both. Therefore, complex ODs with such property can be applied. The ith column can be written as $\mathbf{A}_i \mathbf{b}$ or $\mathbf{B}_i \overline{\mathbf{b}}$ naturally.

As an example, in what follows, the application of the 2×2 Alamouti design [1] in a network with 2 relays, i.e., $R = 2$, is explained [19]. Let $T_1 = T_2 = 2$ and $K = 2$. Let b_1 and b_2 be the two information symbols. They can be modulated by any constellation such as PSK or QAM. Design the vector at the transmitter and the relay processing matrices as

$$\mathbf{b} = \begin{bmatrix} b_1 \\ b_2 \end{bmatrix}, \quad \mathbf{A}_1 = \mathbf{I}_2, \quad \mathbf{B}_1 = \mathbf{0}, \quad \mathbf{A}_2 = \mathbf{0}, \quad \mathbf{B}_2 = \begin{bmatrix} 0 & -1 \\ 1 & 0 \end{bmatrix}.$$

The space-time codeword formed at the receiver has the following form:

$$\mathbf{S} = \begin{bmatrix} b_1 & -b_2^* \\ b_2 & b_1^* \end{bmatrix}, \tag{2.54}$$

whose first column contains b_1, b_2 and second column contains their conjugates exclusively. The space-time code-matrix in (2.54) is the transpose of the originally proposed Alamouti code. It is equivalent to the original one by re-defining $\mathbf{b} = \begin{bmatrix} b_1 & -b_2^* \end{bmatrix}^T$.

Now we consider the application of a general $T \times R$ complex OD with K complex information symbols $b_1, \cdots, b_K$ in a network with R relays. One possible design is as follows [19]. For the first step, the signal vector is designed to be

$$\mathbf{b} = \begin{bmatrix} b_1 & \cdots & b_K & b_1^* & \cdots & b_K^* \end{bmatrix}^t.$$

Thus the first step takes $2K$ time slots, i.e., $T_1 = 2K$. With this first step design, each relay receives a noisy version of the information vector $\begin{bmatrix} b_1 & \cdots & b_K \end{bmatrix}^t$ from the first half of its received signal vector and a noisy version of the conjugate $\begin{bmatrix} b_1 & \cdots & b_K \end{bmatrix}^*$ from the second half of its received signal vector. For the second step, let $\mathbf{B}_i = \mathbf{0}$ for all $i = 1, \cdots, R$ and design the $T_2 \times (2K)$ $\mathbf{A}_i$ matrices based on the structure of the desired complex OD to have the distributed space-time codeword formed at the receiver take the desired format. This is possible since each entry of the code-matrix is a linear combination of b_i's and b_i^*'s. The length of the second step is T_2. The total time slots used for sending the K symbols is $2K + T_2$.

The ML decoding of DSTC is given in (2.51). It should be noted that in general, the noise covariance matrix $\mathbf{R_w}$ is not a multiple of the identity matrix. Thus, the ML decoding cannot be decoupled to symbol-wise decoding. Sphere decoding can still be employed with the decoding formula in (2.44). To have low decoding complexity, an approximation can be made by omitting the matrix $\mathbf{R_w}$ in the ML decoding in (2.51) to obtain the following suboptimal decoding:

$$\arg \min_{\mathbf{b}} \left\| \mathbf{x} - \sqrt{\beta_{\text{gen}}} \mathbf{Sh} \right\|_F^2 . \tag{2.55}$$

This suboptimal decoding can be decomposed into sperate decodings of the information symbols individually. Simulation shows that this suboptimal decoding has very close performance to the optimal ML decoding.

As an example, we consider the following 4×4 complex OD with 3 information symbols b_1, b_2, b_3:

$$\mathbf{S} = \begin{bmatrix} b_1 & b_2 & b_3 & 0 \\ -b_2^* & b_1^* & 0 & b_3 \\ b_3^* & 0 & -b_1^* & b_2 \\ 0 & b_3^* & -b_2^* & -b_1 \end{bmatrix} . \tag{2.56}$$

It can be used in networks with $T_1 = 6$, $T_2 = 4$, and $R = 4$. The transmitted signal vector is design as $\mathbf{b} = \begin{bmatrix} b_1 & b_2 & b_3 & b_1^* & b_2^* & b_3^* \end{bmatrix}^t$. The matrices used at the relays are designed as

$$\mathbf{A}_1 = \frac{4}{3} \begin{bmatrix} 1 & 0 & 0 & 0 & 0 & 0 \\ 0 & 0 & 0 & 0 & -1 & 0 \\ 0 & 0 & 0 & 0 & 0 & 1 \\ 0 & 0 & 0 & 0 & 0 & 0 \end{bmatrix}, \quad \mathbf{A}_2 = \frac{4}{3} \begin{bmatrix} 0 & 1 & 0 & 0 & 0 & 0 \\ 0 & 0 & 0 & 1 & 0 & 0 \\ 0 & 0 & 0 & 0 & 0 & 0 \\ 0 & 0 & 0 & 0 & 0 & 1 \end{bmatrix},$$

$$\mathbf{A}_3 = \frac{4}{3} \begin{bmatrix} 0 & 0 & 1 & 0 & 0 & 0 \\ 0 & 0 & 0 & 0 & 0 & 0 \\ 0 & 0 & 0 & -1 & 0 & 0 \\ 0 & 0 & 0 & 0 & -1 & 0 \end{bmatrix}, \quad \mathbf{A}_4 = \frac{4}{3} \begin{bmatrix} 0 & 0 & 0 & 0 & 0 & 0 \\ 0 & 0 & 1 & 0 & 0 & 0 \\ 0 & 1 & 0 & 0 & 0 & 0 \\ -1 & 0 & 0 & 0 & 0 & 0 \end{bmatrix},$$

$$\mathbf{B}_1 = \mathbf{B}_2 = \mathbf{B}_3 = \mathbf{B}_4 = \mathbf{0}.$$

The coefficient $4/3$ is for the normalization in (2.40). The covariance matrix of the equivalent noise vector for this DSTC design can be calculated from (2.50) to be:

$$\mathbf{R_w} = \left(1 + \frac{4}{3} \alpha \|\mathbf{g}\|_F^2 \right) \mathbf{I}_4 - \frac{4}{3} \alpha \text{diag} \left\{ \left[|g_4|^2, |g_3|^2, |g_2|^2, |g_1|^2 \right] \right\} .$$

Although $\mathbf{R_w}$ is diagonal, its diagonal entries are different. Thus in the ML decoding, the information symbols cannot be decoupled. To have decoupled symbol-wise decoding, we can ignore the quadratic term in (2.51), which is equivalent to approximate $\mathbf{R_w}$ with a multiple of the identity matrix. The resulting decoding is suboptimal.

Although ideal for DSTC, real and complex ODs are sparse and limited in symbol-rate [35]. It was shown in [29] that real ODs whose symbol-rate is 1 and complex

ODs whose symbol rate is 3/4 exist for any dimension. But for $R > 2$, the symbol rates of complex ODs are upper-bounded by 3/4.

2.3.2 Quasi-Orthogonal Code

Quasi-orthogonal design (QOD) was proposed and used in space-time code design in [12, 13, 25, 26, 28, 31], originally for multiple-antenna systems with four transmit antennas, later generalized to other cases. For a QOD, some but not all columns of a code-matrix are orthogonal to each other. There are several different but equivalent ways of constructing a quasi-orthogonal code. Here, we use the one presented in [31, 36].

Let $\mathbf{O}(\mathbf{b_1}, \cdots, \mathbf{b_K})$ be a $T \times M$ complex OD with complex indeterminate variables $b_1, \cdots, b_K$. A $(2T) \times (2M)$ QOD is [31, 36]:

$$\mathbf{S} = \begin{bmatrix} \mathbf{O}(\mathbf{b_1}, \cdots, \mathbf{b_K}) & \mathbf{O}(\mathbf{b_{K+1}}, \cdots, \mathbf{b_{2K}}) \\ \mathbf{O}(\mathbf{b_{K+1}}, \cdots, \mathbf{b_{2K}}) & \mathbf{O}(\mathbf{b_1}, \cdots, \mathbf{b_K}) \end{bmatrix}.$$

It can be shown that $\mathbf{S^*S} = \begin{bmatrix} a\mathbf{I}_M & b\mathbf{I}_M \\ b\mathbf{I}_M & a\mathbf{I}_M \end{bmatrix}$, where $a = \sum_{i=1}^{2K} |b_i|^2$ and $b = \sum_{i=1}^{K}(b_i b_{K+i}^* + b_{K+i} b_i^*)$. Since not all columns of the code matrix are mutually orthogonal, when used in space-time coding, QOD does not have the decoupled symbol-wise decoding. However, due to its special structure, it has pairwise symbol decoding, i.e., information symbols can be decoded in joint pairs. Thus it is still highly superior in decoding complexity to the joint decoding of all information symbols, especially for systems with high dimensions and high transmission rates. From the definition of QOD, it can be seen that similar to OD, QOD can be used for DSTC straightforwardly.

For the special case that both $\mathbf{O_1}$ and $\mathbf{O_2}$ are Alamouti design, the 4×4 QOD has the following structure:

$$\begin{bmatrix} b_1 & -b_2^* & b_3 & -b_4^* \\ b_2 & b_1^* & b_4 & b_3^* \\ b_3 & -b_4^* & b_1 & -b_2^* \\ b_4 & b_3^* & b_2 & b_1^* \end{bmatrix}. \tag{2.57}$$

In what follows, we demonstrate the use of this QOD for DSTC in networks with 4 relays, i.e., $R = 4$. Let $T_1 = T_2 = 4$ and $K = 4$. Let b_1, b_2, b_3, b_4 be the 4 information symbols. Design the vector sent by the transmitter and the relay transformation matrices as

$$\mathbf{b} = \begin{bmatrix} b_1 \\ b_2 \\ b_3 \\ b_4 \end{bmatrix}, \quad \mathbf{A_1} = \mathbf{I_4}, \quad \mathbf{A_2} = 0, \quad \mathbf{A_3} = \begin{bmatrix} 0 & 0 & 1 & 0 \\ 0 & 0 & 0 & 1 \\ 1 & 0 & 0 & 0 \\ 0 & 1 & 0 & 0 \end{bmatrix}, \quad \mathbf{A_4} = 0,$$

$$B_1 = 0, \quad B_2 = \begin{bmatrix} 0 & -1 & 0 & 0 \\ 1 & 0 & 0 & 0 \\ 0 & 0 & 0 & -1 \\ 0 & 0 & 1 & 0 \end{bmatrix}, \quad B_3 = 0, \quad B_4 = \begin{bmatrix} 0 & 0 & 0 & -1 \\ 0 & 0 & 1 & 0 \\ 0 & -1 & 0 & 0 \\ 1 & 0 & 0 & 0 \end{bmatrix}.$$

With this design, the space-time codeword formed at the receiver has the quasi-orthogonal structure in (2.57).

2.3.3 Algebraic Code

An algebraic family of distributed space-time code was proposed in [22], which can be used for networks with any relays and $T_1 = T_2 = R$. Let b be the information signal sent by the transmitter. Elements of b can be designed as PSK, QAM, or other modulations. Define

$$G \triangleq \begin{bmatrix} 0 & 0 & 0 & \cdots & i \\ 1 & 0 & 0 & \cdots & 0 \\ 0 & 1 & 0 & \cdots & 0 \\ \vdots & \vdots & \ddots & \ddots & \vdots \\ 0 & 0 & \cdots & 1 & 0 \end{bmatrix},$$

which is a $T \times T$ matrix. The linear transformation matrix at the relays are designed as $A_i = G^{i-1}$ and $B_i = 0$ for $i = 1, \cdots, R$. The distributed space-time codeword formed at the receiver has the following structure

$$S = \begin{bmatrix} b & Gb & \cdots & G^{R-1}b \end{bmatrix}.$$

It has been proved in [22] using involved cyclotomic algebra that algebraic code achieves full diversity. More designs of algebraic code can be found in [23, 24].

2.4 Simulation on Error Probability

In this section, simulated error probabilities of DSTC are demonstrated. Comparison with corresponding space-time coding for multiple-antenna systems is also made.

In all simulations, the channels and noises are generated independently following $\mathcal{CN}(0, 1)$. For the transmit power, we set $P_s = RP_r$, which follows the optimal power allocation derived in Sect. 2.2.3. For each transmission block, an information vector b is generated randomly. The ML decoding is conducted. If the decoded vector, denoted as $\hat{b}$, is different to b, a block error occurs. A large number of transmission blocks are simulated. Distinct and independent channel realizations are used for

different transmission blocks. The average block error rate is approximated as the
ratio of the total number of block errors to the total number of blocks.

First, networks with $1, 2, 3, 4$ relays, i.e., $R = 1, 2, 3, 4$ are considered to show
the achievable diversity orders of networks with different numbers of relays. In this
simulation experiment, the basic DSTC scheme explained in Sect. 2.2.2 is used.
The transmission time for each step T is chosen to be the same as R. Entries of
the information vector $\mathbf{b}$ are generated as i.i.d. binary phase-shift keying (BPSK)
signals. Thus the overall transmission rate is 1/2 bit per transmission. As to the relay
transformation matrices, for each transmission block, a distinct set of unitary matrices
$\mathbf{A}_1, \cdots, \mathbf{A}_R$ are generated according to the isotropic distribution. This follows the
random code [17, 19] idea. When the number of blocks is large enough, the simulated
performance approximates the average performance of all possible DSTCs. The
use of random code is to eliminate the affect of the code design and to focus on
the diversity order only. In Fig. 2.4, block error rates of DSTC v.s. P_s are shown.
The diversity orders can be observed from the slopes of the curves. We can see that
the diversity order of DSTC is slightly less than R for a relay network with R relays.
As P_s increases, the diversity order increases and approaches the corresponding R
value. This complies with the diversity order result in Sect. 2.2.3.

In the next experiment, a network with $T = R = 2$ is considered. The block
error rates of two code designs, Alamouti code and Algebraic code, are demonstrated. QPSK is used for both information symbols in the information vector $\mathbf{b}$.
So the transmission rate is 1 bit per transmission. The performance of a multiple-
antenna system with 2 transmit antennas and 1 receive antenna with Alamouti code

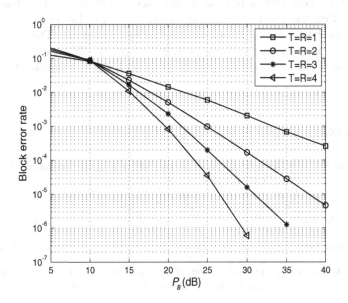

Fig. 2.4 Performance of DSTC in networks with $R = 1, 2, 3, 4$ relays, random code, and BPSK

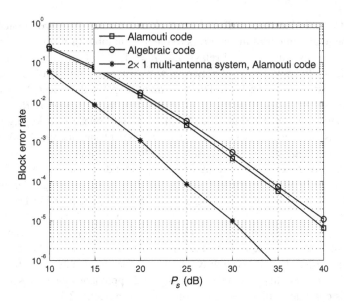

Fig. 2.5 Performance of DSTC in a network with $T = R = 2$, QPSK; and comparison with multiple-antenna system with 2 transmit antennas and 1 receive antenna

is also shown for comparison. For the multiple-antenna system, the transmit power is set to be P_s. From Fig. 2.5, it can be seen that for the relay network, the diversity order of both codes approaches 2 as P increases. Alamouti code has slight lower error rate. Compared with the multiple-antenna system under Alamouti code, the network has about the same diversity order, but higher error rate. The performance difference of the two systems is about 9 dB for high P_s. This is due to the noise propagation at the relay and also the two stages of fading channels in the relay network.

References

1. Alamouti SM (1998) A simple transmitter diversity scheme for wireless communications. IEEE J on Selected Areas in Communications, 16:1451–1458.
2. Barbero LG and Thompson JS (2008) Fixing the complexity of the sphere decoder for MIMO detection. IEEE T Wireless Communications, 7:2131–2142.
3. Damen O, Chkeif A, and Belfiore JC (2000) Lattice code decoder for space-time codes. IEEE Communications L, 4:161–163.
4. Gradshteyn IS and Ryzhik IM (2000) Table of Integrals, Series and Products. Academic Press, 6nd ed.
5. Elia P, Kittipiyakul S, and Javidi T (2007) Cooperative diversity schemes for asynchronous wireless networks. Wireless Personal Communications, 43: 3–12, doi:10.1007/s11277-006-9242-3.
6. Guo X and Xia XG (2008) A distributed space-time coding in asynchronous wireless relay networks. IEEE T on Wireless Communications, 7:1812–1816.

7. Hassibi B and Hochwald B (2002) High-rate codes that are linear in space and time. IEEE T on Information Theory, 48:1804–1824.

8. Hassibi B and Vikalo H (2005) On the sphere-decoding algorithm I. Expected complexity. IEEE T Signal Processing, 8:2806–2818.

9. Hochwald BM and Marzetta TL (2000) Unitary space-time modulation for multiple-antenna communication in Rayleigh flat-fading. IEEE T on Information Theory, 46:543–564.

10. Hochwald B and Sweldens W (1999) Differential unitary space-time modulation. IEEE T on Communications, 48:2041–2052.

11. Hughes B (2000) Differential space-time modulation. IEEE T on Information Theory, 46:2567–2578.

12. Jafarkhani H (2001) A quasi-orthogonal space-time block codes. IEEE T on Communications, 49:1–4.

13. Jafarkhani H (2005) Space-Time Coding: Theory and Practice. Cambridge Academic Press.

14. Jing Y (2010) Combination of MRC and distributed space-time coding in networks with multiple-antenna relays. IEEE T on Wireless Communications, 9:2550–2559.

15. Jing Y (2004) Space-Time Code Design and Its Applications in Wireless Networks. Ph.D. Thesis, California Institute of Technology.

16. Jing Y and Hassibi B (2004) Wireless network, diversity, and space-time codes. Information Theory, Workshop, 2004.

17. Jing Y and Hassibi B (2006) Distributed space-time coding in wireless relay networks. IEEE T on Wireless Communications, 5:3524–3536.

18. Jing Y and Hassibi B (2008) Cooperative diversity in wireless relay networks with multiple-antenna nodes. EURASIP J on Advanced Signal Process, doi:10.1155/2008/254573.

19. Jing Y and Jafarkhani H (2007) Using orthogonal and quasi-orthogonal designs in wireless relay networks. IEEE T on Information Theory, 53:4106–4118.

20. Kiran T and Rajan BS (2006) Distributed space-time codes with reduced decoding complexity. IEEE International Symposium on Information Theory, 542–546.

21. Li Y, Zhang W, and Xia X-G (2009) Distributive high-rate space-frequency codes achieving full cooperative and multipath diversities for asynchronous cooperative communications. IEEE T on Vehicular Technology, 58:207–217.

22. Oggier F and Hassibi B (2006) An algebraic family of distributed space-time codes for wireless relay networks. IEEE International Symposium on Information Theory, 538–541.

23. Oggier F and Hassibi B (2008) An algebraic coding scheme for Wireless Relay networks with Multiple-Antenna Nodes. IEEE T Signal Processing, 56:2957–2966.

24. Oggier F and Hassibi B (2008) Code design for multihop wireless relay networks. EURASIP J Advances Signal Processing, 2008. doi:10.1155/2008/457307.

25. Papadias CB and Foschini GJ (2011) A space-time coding approach for systems employing four transmit antennas. International C on on Acoustics, Speech, and Signal Processing, 4:2481–2484.

26. Papadias CB and Foschini GJ (2003) Capacity-approaching space-time codes for systems employing four transmitter antennas. IEEE T Information Theory, 49:726–732.

27. Rajan GS and Rajan BS (2007) Algebraic distributed space-time codes with low ML decoding complexity. IEEE International Symposium on Information Theory, 1516–1520.

28. Su W and Xia X-G (2004) Signal constellations for quasi-orthogonal space-time block codes with full diversity. IEEE T on Information Theory, 50:2331–2347.

29. Tarokh V, Jafarkhani H, and Calderbank AR (1999) Space-time block codes from orthogonal designs. IEEE T on Information Theory, 45:1456–1467.

30. Tarokh V, Seshadri N, and Calderbank AR (1998) Space-time codes for high data rate wireless communication: Performance criterion and code construction. IEEE T on Information Theory, 44:744–765.

31. Tirkkonen O, Boariu A, and Hottinen A (2000) Minimal non-orthogonality rate 1 space-time block code for 3-Tx antennas. IEEE 6th Int. Symp. Spread-Spectrum Tech. Appl., 429432.

32. Van Trees HL (1998) Detection, Estimation, and Modulation Theory-Part I. New York: Wiley.

33. Vikalo H and Hassibi B (2005) On the sphere-decoding algorithm II. Generalizations, second-order statistics, and applications to communications. IEEE T Signal Processing, 8:2819–2834.
34. Viterbo E and Boutros J (1999) A universal lattice code decoder for fading channels. IEEE T Information Theory, 5:1639–1642.
35. Wang H and Xia X-G (2003) Upper bounds of rates of complex orthogonal space-time block codes. IEEE T on Information Theory, 49:2788–2796.
36. Wang H and Xia X-G (2005) On optimal quasi-orthogonal space-time block codes with minimum decoding complexity. IEEE International S on Information Theory, 1168–1172.
37. Yi Z, Ju M, Song H-K, and Kim I-M (2011) BER analysis of distributed Alamouti's code with CSI-assiated relays. IEEE T Wireless Communications, 10:1199–1211.
38. Yi Z and Kim DI (2007) Single-Symbol ML decodable distributed STBCs for cooperative networks. IEEE T Information Theory, 53:2977–2985.
39. Yi Z and Kim I-M (2009) Approximate BER expressions of distributed Alamouti's code in dissimilar cooperative networks with blind relays. IEEE T Communications, 57:3571–3578.
40. Yi Z, Kim I-M, and Kim DI (2011) Symbol rate upper bound of distributed STBC with channel phase information. IEEE T Wireless Communications, 10:745–750.
41. Zummo SD and Al-Semari SA (2000) A tight bound on the error probability of space-time codes for rapid fading channels. IEEE, Wireless Communications and Networking Conference, 1086–1089.
42. Uysal M and Georghiades CN (2000) Error performance analysis of space-time codes over Rayleigh fading channels. IEEE Vehicular Technology Conference- fall, 5:2285–2290.

Chapter 3
Distributed Space-Time Coding for Multiple-Antenna Multiple-Relay Network

Abstract In the previous chapter, distributed space-time coding (DSTC) is introduced for *single-antenna multiple-relay network*, where there are multiple relays in the network but every node (the transmitter, the receiver, or a relay) is equipped with a single antenna. In this chapter, DSTC is used for *multiple-antenna multiple-relay network*, where each node in the network can be equipped with multiple antennas. At first, the multiple-antenna multiple-relay network model is specified in Sect. 3.1. Then the DSTC scheme, its diversity order analysis, and code design are demonstrated in Sect. 3.2. After that, the combination of DSTC and maximum-ratio combining (MRC) is explained in Sect. 3.3. Finally, simulated error probabilities of some multiple-antenna multiple-relay networks are shown.

3.1 Multiple-Antenna Multiple-Relay Network Model

Consider a wireless network with one transmitter, one receiver, and multiple relays. As shown in Fig. 3.1. the general multiple-antenna case is considered, where the transmitter has M transmit antennas, the receiver has N receive antennas, and the relays have a total of R antennas, which can be used for both transmission or reception. Denote the channel vector from the M antennas of the transmitter to the ith relay antenna as $\mathbf{f}_i = \begin{bmatrix} f_{1i} & \cdots & f_{Mi} \end{bmatrix}^t$, where f_{mi} is the channel between the mth transmit antenna and the ith relay antenna. Denote the channel vector from the ith relay antenna to the N antennas at the receiver as $\mathbf{g}_i = \begin{bmatrix} g_{i1} & \cdots & g_{iN} \end{bmatrix}$, where g_{in} is the channel between the ith relay antenna and the nth receive antenna. Define

$$\mathbf{f} \triangleq \begin{bmatrix} \mathbf{f}_1 \\ \vdots \\ \mathbf{f}_R \end{bmatrix}, \quad \mathbf{G} \triangleq \begin{bmatrix} \mathbf{g}_1 \\ \vdots \\ \mathbf{g}_R \end{bmatrix}, \tag{3.1}$$

which are the $MR \times 1$ transmitter-relay channel vector and the $R \times N$ relay-receiver channel matrix. There is no direct link between the transmitter and the receiver. Independent Rayleigh flat-fading channels are assumed, i.e., f_{mi}'s and g_{in}'s are i.i.d. following $\mathcal{CN}(0, 1)$. Further, a block-fading model with coherence interval T is assumed. Assume that the receiver has global and perfect channel state information (CSI), i.e., the receiver knows all f_{mi}'s and g_{in}'s. But the transmitter and the relays have no CSI.

3.2 DSTC for Multiple-Antenna Multiple-Relay Network

In this section, the distributed space-time coding (DSTC) protocol for multiple-antenna relay network is introduced. The diversity order of DSTC is also derived along with discussions on DSTC code design.

3.2.1 DSTC Protocol

As explained in Chap. 2, DSTC is a two-step transmission scheme. In the first step, the transmitter sends information to the relays. The relays then linearly process their received signals and forward information to the receiver in the second step. The details are as follows.

Step 1: Transmission from the Transmitter to the Relays

The first step is the transmission from the transmitter to the relays. Information bits are encoded into a $T \times M$ matrix **B**, normalized to

Fig. 3.1 Multiple-antenna multiple-relay network

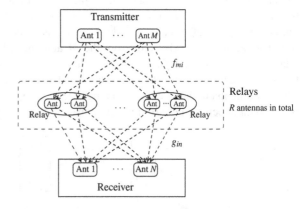

$$\mathbb{E}\left\{\text{tr}\left(\mathbf{B}^*\mathbf{B}\right)\right\} = M. \tag{3.2}$$

The transmitter sends $\sqrt{P_sT/M}\mathbf{B}$. That is, the mth transmit antenna of the transmitter sends the mth column of $\mathbf{B}$ multiplied with the power coefficient $\sqrt{P_sT/M}$. With the normalization in (3.2), the average power used at the transmitter is P_s per transmission. The same as the notation in Sect. 2.2.2, the received signal vector and the noise vector at the ith relay antenna are denoted as $\mathbf{r}_i$ and $\mathbf{n}_{r,i}$, respectively. Thus,

$$\mathbf{r}_i = \sqrt{\frac{P_sT}{M}}\mathbf{B}\mathbf{f}_i + \mathbf{n}_{r,i}. \tag{3.3}$$

The first step takes T time slots.

Relay Processing

After receiving $\mathbf{r}_i$, the ith relay antenna linearly processes $\mathbf{r}_i$ using a pre-determined $T \times T$ matrix $\mathbf{A}_i$ to obtain $\mathbf{t}_i$ as:

$$\mathbf{t}_i = \sqrt{\alpha}\mathbf{A}_i\mathbf{r}_i, \tag{3.4}$$

where $\alpha \triangleq \frac{P_r}{1+P_s}$. This definition of α is the same as the one in (2.16).

In general, if a relay is equipped with multiple antennas, the relay can jointly process the received signal vectors of all its antennas. Thus the transmitted signal vector of any antenna of the relay can depend on the received signal vectors of all antennas of the same relay. But this section is only concerned with the simple design that the transmitted signal of each relay antenna depends on the received signal of that antenna only, regardless that it may have co-located antennas. As will be seen later, this simple design achieves the optimal diversity order for asymptotically high transmit power. A design that considers joint signal processing among co-located antennas on the same relay will be explained in Sect. 3.3.

Step 2: Transmission from the Relays to the Receiver

In the second step, the ith relay antenna sends $\mathbf{t}_i$. With the normalization in (3.2) and (3.4), it can be shown that the average transmit power of each relay antenna is P_r per transmission. The relays are assumed to be synchronized so the signals they sent arrive the receiver at the same time. Denote the $T \times N$ noise matrix at the receiver as $\mathbf{N}_d$. The $T \times N$ received signal matrix, denoted as $\mathbf{X}$, can be represented as

$$\mathbf{X} = \begin{bmatrix} \mathbf{t}_1 & \cdots & \mathbf{t}_R \end{bmatrix}\mathbf{G} + \mathbf{N}_d, \tag{3.5}$$

where $\mathbf{G}$ is defined in (3.1).

Transceiver Equation

By using (3.3) and (3.4) in (3.5), after some straightforward calculations, the system equation of DSTC in multiple-antenna multiple-relay network can be written as

$$\mathbf{X} = \sqrt{\beta}\mathbf{SH} + \mathbf{W}, \tag{3.6}$$

where

$$\beta \triangleq \frac{P_s P_r T}{(P_s + 1)M}, \tag{3.7}$$

$$\mathbf{S} \triangleq \begin{bmatrix} \mathbf{A}_1\mathbf{B} \cdots \mathbf{A}_R\mathbf{B} \end{bmatrix}, \tag{3.8}$$

$$\mathbf{H} \triangleq \begin{bmatrix} \mathbf{f}_1\mathbf{g}_1 \\ \vdots \\ \mathbf{f}_R\mathbf{g}_R \end{bmatrix} = \text{diag}\{\mathbf{f}_1, \cdots, \mathbf{f}_R\}\mathbf{G}, \tag{3.9}$$

$$\mathbf{W} \triangleq \sqrt{\alpha}\begin{bmatrix} \mathbf{A}_1\mathbf{n}_{r,1} \cdots \mathbf{A}_R\mathbf{n}_{r,R} \end{bmatrix}\mathbf{G} + \mathbf{N}_d. \tag{3.10}$$

Recall that for single-antenna relay network, β is defined in (2.16). Here with slight abuse of notation but for the consistency of the presentation, we redefine β in (3.7) for multiple-antenna relay network. When $M = 1$, the two definitions coincide. $\mathbf{S}$, which is $T \times MR$, is the distributed space-time codeword formed at the receiver. $\mathbf{H}$ is the end-to-end channel matrix. Since $\mathbf{f}_i$ is $M \times 1$ and $\mathbf{g}_i$ is $1 \times N$, it is $RM \times N$. Note that the rth block of $\mathbf{H}$, which is $\mathbf{f}_r\mathbf{g}_r$ is $M \times N$. It is the end-to-end channel matrix from the transmitter to the receiver via the rth relay antenna. $\mathbf{W}$ is the equivalent noise matrix. It is $T \times N$.

An equivalent form of the transceiver equation can be obtained by stacking the columns of $\mathbf{X}$ into one column vector to be:

$$\text{vec}(\mathbf{X}) = \sqrt{\beta}(\mathbf{G}^t \otimes \mathbf{I}_T)\tilde{\mathbf{S}}\mathbf{f} + \text{vec}(\mathbf{W}), \tag{3.11}$$

where

$$\tilde{\mathbf{S}} \triangleq \text{diag}\{\mathbf{A}_1\mathbf{B}, \cdots, \mathbf{A}_R\mathbf{B}\}.$$

It is an equivalent format for the distributed space-time codeword. Let

$$\mathbf{f}_{\text{diag}} \triangleq \text{diag}\{\mathbf{f}_1, \cdots, \mathbf{f}_R\}. \tag{3.12}$$

Another equivalent form is

$$\mathbf{X} = \sqrt{\beta}\mathbf{S}\mathbf{f}_{\text{diag}}\mathbf{G} + \mathbf{W}. \tag{3.13}$$

3.2.2 ML Decoding of DSTC

Now we derive the maximum likelihood (ML) decoding of DSTC. All noises are assumed to be i.i.d. $\mathcal{CN}(0, 1)$. To derive the ML decoding, the likelihood function, which is the conditional probability density function (PDF) of $\mathbf{X}|\mathbf{B}$ is needed. From (3.6)–(3.10), the (t, n)-th entry of $\mathbf{X}$, denoted as x_{tn}, equals

$$x_{tn} = \sqrt{\beta} \sum_{i=1}^{R} \sum_{m=1}^{M} \sum_{\tau=1}^{T} f_{mi} g_{in} a_{i,t\tau} b_{\tau m} + \sqrt{\alpha} \sum_{i=1}^{R} \sum_{\tau=1}^{T} g_{in} a_{i,t\tau} n_{r,i,\tau} + n_{d,tn},$$

where $a_{i,t\tau}$ is the (t, τ)th entry of $\mathbf{A}_i$, $b_{\tau m}$ is the (τ, m)th entry of $\mathbf{B}$, $n_{r,i,\tau}$ is the τth entry of $\mathbf{n}_{r,i}$, and $n_{d,tn}$ is the (t, n)th entry of $\mathbf{N}_d$. With full CSI at the receiver,

$$\mathbb{E}(x_{tn}) = \sqrt{\beta} \sum_{i=1}^{R} \sum_{m=1}^{M} \sum_{\tau=1}^{T} f_{mi} g_{in} a_{i,t\tau} b_{\tau m}.$$

Therefore, the mean of the t-th row of $\mathbf{X}$ is $\sqrt{\beta}[\mathbf{B}]_t \mathbf{H}$ with $[\mathbf{B}]_t$ the t-th row of $\mathbf{B}$. Since $\mathbf{n}_{r,i}$'s, $\mathbf{N}_d$, and $\mathbf{B}$ are independent, the covariance of $x_{t_1 n_1}$ and $x_{t_2 n_2}$ can be calculated as follows:

$$\text{Cov}\left(x_{t_1 n_1}, x_{t_2 n_2}\right)$$
$$= \mathbb{E}\left\{\left(x_{t_1 n_1} - \mathbb{E}\{x_{t_1 n_1}\}\right) \overline{\left(x_{t_2 n_2} - \mathbb{E}\{x_{t_2 n_2}\}\right)}\right\}$$
$$= \alpha \sum_{i_1=1}^{R} \sum_{\tau_1=1}^{T} \sum_{i_2=1}^{R} \sum_{\tau_2=1}^{T} \mathbb{E}\{g_{i_1 n_1} a_{i_1, t_1 \tau_1} n_{r,i_1,\tau_1} \overline{g_{i_2 n_2} a_{i_2, t_2 \tau_2} n_{r,i_2,\tau_2}}\} + \mathbb{E}\{n_{d,t_1 n_1} \overline{n_{d,t_2 n_2}}\}$$
$$= \alpha \sum_{i=1}^{R} \left(\sum_{\tau=1}^{T} a_{i,t_1 \tau} \overline{a_{i,t_2 \tau}}\right) g_{in_1} \overline{g_{in_2}} + \delta_{n_1 n_2} \delta_{t_1 t_2}$$
$$= \delta_{t_1 t_2} \left(\alpha \sum_{i=1}^{R} g_{in_1} \overline{g_{in_2}} + \delta_{n_1 n_2}\right)$$
$$= \delta_{t_1 t_2} \left(\alpha \left[g_{1n_1} \cdots g_{Rn_1}\right] \begin{bmatrix} \bar{g}_{1n_2} \\ \vdots \\ \bar{g}_{Rn_2} \end{bmatrix} + \delta_{n_1 n_2}\right).$$

The fourth equality is true since $\mathbf{A}_i$ is unitary. The previous derivations show that the rows of $\mathbf{X}$ are uncorrelated since the covariance of $x_{t_1 n_1}$ and $x_{t_2 n_2}$ is zero when $t_1 \neq t_2$. Since the noises are Gaussian, each row of $\mathbf{X}$ is a Gaussian random vector. Thus, rows of $\mathbf{X}$ are independent.

Define

$$\mathbf{R_W} \triangleq \mathbf{I}_N + \alpha \mathbf{G}^* \mathbf{G}, \tag{3.14}$$

which is a positive definite matrix. The above calculation shows that the covariance matrix of the tth row of $\mathbf{X}$ is $\overline{\mathbf{R_W}}$. Therefore, the PDF of $[\mathbf{X}]_t | \mathbf{B}$ is

$$p([\mathbf{X}]_t | \mathbf{B}) = \left(\pi^N \det \overline{\mathbf{R_W}} \right)^{-1} e^{-\text{tr}\left\{ \overline{[\mathbf{X} - \sqrt{\beta}\mathbf{SH}]_t}\, \overline{\mathbf{R_W}^{-1}} [\mathbf{X} - \sqrt{\beta}\mathbf{SH}]_t^t \right\}}$$

$$= \left(\pi^N \det \mathbf{R_W} \right)^{-1} e^{-\text{tr}\left\{ [\mathbf{X} - \sqrt{\beta}\mathbf{SH}]_t \mathbf{R_W}^{-1} [\mathbf{X} - \sqrt{\beta}\mathbf{SH}]_t^* \right\}}.$$

As rows of $\mathbf{X}$ are independent, it follows that

$$p(\mathbf{X} | \mathbf{B}) = \prod_{t=1}^{T} p([\mathbf{X}]_t | \mathbf{B}) = \left(\pi^N \det \mathbf{R_W} \right)^{-T} e^{-\text{tr}(\mathbf{X} - \sqrt{\beta}\mathbf{SH})\mathbf{R_W}^{-1}(\mathbf{X} - \sqrt{\beta}\mathbf{SH})^*}.$$

$$(3.15)$$

In view of the above investigation, we see that for a relay network with multiple antennas at the receiver, the rows of $\mathbf{X}$ are independent. The columns of $\mathbf{X}$, however, are not independent. This can be seen by realizing that the covariance matrix $\mathbf{R_W}$ is not diagonal in general. That is, the received signals at the same *time* but different receive *antennas* are not independent, whereas the received signals at different *times* are independent.

With $p(\mathbf{X} | \mathbf{B})$ in hand, the ML decoding of DSTC can be derived to be:

$$\arg \max_{\mathbf{B}} p(\mathbf{X} | \mathbf{B}) = \arg \min_{\mathbf{B}} \text{tr} \left\{ \left(\mathbf{X} - \sqrt{\beta}\mathbf{SH} \right) \mathbf{R_W}^{-1} \left(\mathbf{X} - \sqrt{\beta}\mathbf{SH} \right)^* \right\}. \quad (3.16)$$

An equivalent form of the ML decoding is:

$$\arg \min_{\mathbf{B}} \text{tr} \left\{ \left(\mathbf{X} - \sqrt{\beta}\mathbf{Sf}_{\text{diag}}\mathbf{G} \right) \mathbf{R_W}^{-1} \left(\mathbf{X} - \sqrt{\beta}\mathbf{Sf}_{\text{diag}}\mathbf{G} \right)^* \right\}. \quad (3.17)$$

3.2.3 Diversity Order Analysis

Previously, the transmission protocol, the transceiver equation, and ML decoding for DSTC in multiple-antenna multiple-relay network have been explained. In this subsection, the performance of DSTC is investigated, via calculating its PEP and diversity order.

Consider two different information matrices $\mathbf{B}_k$ and $\mathbf{B}_l$. The distributed space-time codewords corresponding to them are denoted as $\mathbf{S}_k$ and $\mathbf{S}_l$, respectively. Thus,

$$\mathbf{S}_k \triangleq \left[\mathbf{A}_1 \mathbf{B}_k \cdots \mathbf{A}_R \mathbf{B}_k \right], \quad \mathbf{S}_l \triangleq \left[\mathbf{A}_1 \mathbf{B}_l \cdots \mathbf{A}_R \mathbf{B}_l \right].$$

Recall the definition in (2.24):

$$\mathbf{M}_{kl} \triangleq (\mathbf{S}_k - \mathbf{S}_l)^* (\mathbf{S}_k - \mathbf{S}_l).$$

With the ML decoding in (2.51), the PEP of the multiple-antenna multiple-relay network under DSTC can be analyzed similarly to that in Sect. 2.2.3 to obtain the following Chernoff bound:

$$\mathbb{P}(\mathbf{B}_k \to \mathbf{B}_l) \leq \mathbb{E}_{\mathbf{f},\mathbf{G}} \left\{ e^{-\frac{\beta}{4} \text{tr}\left(M_{kl} \mathbf{H} \mathbf{R}_{\mathbf{W}}^{-1} \mathbf{H}^*\right)} \right\}. \tag{3.18}$$

Power Allocation Between the Transmitter and the Relays

Before further calculating the PEP bound in (3.18), we derive the power allocation between the transmitter and the relays, or equivalently, between the two transmission steps. Assume that the total power used in the whole network is P. Thus $P = P_s + RP_r$. Notice that when $R \to \infty$, according to the law of large numbers, the off-diagonal entries of $\mathbf{G}^*\mathbf{G}/R$ approaches zero while the diagonal entries approach 1 with probability 1. Thus, for large R, it follows that $\mathbf{G}^*\mathbf{G}/R \approx \mathbf{I}_N$ and $\mathbf{R}_{\mathbf{W}} \approx (1 + \alpha R)\mathbf{I}_N$. With this approximation, minimizing the PEP upper bound in right-hand-side of (3.18) is equivalent to maximizing $\frac{\beta}{4(1+\alpha R)} = \frac{P_s P_r T}{4M(1+P_s+RP_r)}$. This is exactly the same power allocation problem in Sect. 2.2.3. Therefore, it is concluded that the optimum allocation is:

$$P_s = \frac{P}{2} \quad \text{and} \quad P_r = \frac{P}{2R}. \tag{3.19}$$

This results says that the transmitter uses half the total power and the relays share the other half, with the power of each relay antenna being $P/2/R$. The transmit powers of the two transmission steps are the same.

Diversity Order Result

For the diversity order analysis, the PEP behaviour with respect to P for large P needs to be investigated. The main difficulty in the PEP analysis lies in the fact that the noise covariance matrix $\mathbf{R}_{\mathbf{W}}$ is not diagonal. To conquer this difficulty, we derive an upper bound on $\mathbf{R}_{\mathbf{W}}$ and use it to further bound the PEP from the above. Since $\mathbf{R}_{\mathbf{W}}$ is positive definite, we have

$$\mathbf{R}_{\mathbf{W}} \preceq [\text{tr}(\mathbf{R}_{\mathbf{W}})]\mathbf{I}_N = \sum_{n=1}^{N}\left(1 + \alpha \sum_{i=1}^{R} |g_{in}|^2\right)\mathbf{I}_N = \left(N + \alpha\|\mathbf{G}\|_F^2\right)\mathbf{I}_N. \tag{3.20}$$

Therefore, when $P \gg 1$, from (3.18) and by using the power allocation given in (3.19), we have

$$\mathbb{P}(\mathbf{B}_k \to \mathbf{B}_l)$$
$$\leq \mathbb{E}_{\mathbf{f},\mathbf{G}} e^{-\frac{PT}{8NR}\left(1+\frac{2}{P}+\frac{1}{NR}\|\mathbf{G}\|_F^2\right)^{-1} \text{tr}(\mathbf{H}^* M_{kl} \mathbf{H})}$$

$$\leq \mathbb{E}_{\mathbf{f},\mathbf{G}} e^{-\frac{PT}{8NR}\left(1+\frac{1}{NR}\|\mathbf{G}\|_F^2\right)^{-1} \operatorname{tr}(\mathbf{H}^*\mathbf{M}_{kl}\mathbf{H})} + \text{lower order terms in } P$$

$$= \mathbb{E}_{\mathbf{f},\mathbf{G}} e^{-\frac{PT}{8NR}\left(1+\frac{1}{NR}\|\mathbf{G}\|_F^2\right)^{-1} \mathbf{f}^*\left[\sum_{n=1}^{N} \mathcal{G}_n^* \mathbf{M}_{kl} \mathcal{G}_n\right]\mathbf{f}} + \text{lower order terms in } P,$$

where $\mathcal{G}_n \triangleq \operatorname{diag}\{[g_{1n} \cdots g_{Rn}]\} \otimes \mathbf{I}_M$. Since $\mathbf{f}$ is Gaussian distributed whose mean is zero and whose variance is $\mathbf{I}_{RM}$, by calculating the average over $\mathbf{f}$, we have

$$\mathbb{P}(\mathbf{B}_k \to \mathbf{B}_l) \leq \mathbb{E}_{\mathbf{G}} \det{}^{-1}\left[\mathbf{I}_{RM} + \frac{PT}{8NR}\left(1+\frac{1}{NR}\|\mathbf{G}\|_F^2\right)^{-1}\sum_{n=1}^{N} \mathcal{G}_n^* \mathbf{M}_{kl} \mathcal{G}_n\right]$$

$$+ \text{lower order term in } P. \tag{3.21}$$

Denote the minimum singular value of $\mathbf{M}_{kl}$ by $\sigma_{\min}^2$. Assume that $T \geq MR$ and the distributed space-time block code has full diversity. Thus, $\sigma_{\min}^2 > 0$. Since only the diversity order is concerned in this section, for the simplicity of the representation, we only keep the highest order term of P in the Chernoff bound and neglect the lower order terms. From (3.21), we have

$$\mathbb{P}(\mathbf{B}_k \to \mathbf{B}_l) \lesssim \mathbb{E}_{\mathbf{G}} \det{}^{-1}\left[\mathbf{I}_{RM} + \frac{PT\sigma_{\min}^2}{8NR}\left(1+\frac{1}{NR}\|\mathbf{G}\|_F^2\right)^{-1}\sum_{n=1}^{N} \mathcal{G}_n^* \mathcal{G}_n\right]$$

$$= \mathbb{E}_{\mathbf{G}} \prod_{i=1}^{R}\left(1 + \frac{PT\sigma_{\min}^2}{8NR}\frac{\|\mathbf{g}_i\|_F^2}{1+\frac{1}{NR}\|\mathbf{G}\|_F^2}\right)^{-M}. \tag{3.22}$$

Further calculating the average over $\mathbf{G}$, the following PEP upper bound can be derived:

$$\mathbb{P}(\mathbf{B}_k \to \mathbf{B}_l) \leq c_{\text{DSTC-mul}} \cdot \begin{cases} \left[\frac{M}{N(M-N)}\right]^R P^{-NR} & \text{if } M > N \\ \left(1+\frac{1}{N}\right)^R P^{-MR}\log_e^R P & \text{if } M = N \\ \left[\frac{1}{N}+(N-M-1)!\right]^R P^{-MR} & \text{if } M < N \end{cases}, \tag{3.23}$$

where $c_{\text{DSTC-mul}} \triangleq \frac{1}{(N-1)!^R}\left(\frac{8NR}{T\sigma_{\min}^2}\right)^{\min\{M,N\}R}$.

The derivation of (3.23) from (3.22) is tedious and can be found in [4]. In obtaining (3.23), an upper bound on $\mathbf{R_W}$, given in (3.20), is used. This bound is loose, so the PEP bound in (3.23), although provides the desired diversity order result, is loose in evaluating the error probability of the network. A tighter bound is available in [4].

Based on (3.23), by noticing that

$$P^{-MR}\log_e^R P = P^{-MR+R\frac{\log_e \log_e P}{\log_e P}},$$

the following theorem on diversity order can be obtained.

Theorem 3.1. *For a relay network with M transmit antennas, N receive antennas, and a total of R antennas at the relays, assume that $T \geq MR$ and the distributed space-time code is fully diverse. An achievable diversity order of DSTC is*

$$d_{DSTC\text{-}mul} = \begin{cases} \min\{M, N\}R & \text{if } M \neq N \\ MR - R\frac{\log_e \log_e P}{\log_e P} & \text{if } M = N \end{cases}. \tag{3.24}$$

For any scheme with a two-step protocol (where Step 1 is the transmission from the transmitter to the relays and Step 2 is the transmission from the relays to the receiver), the performance is limited by the worse of the two transmission steps. The maximum diversity order of the first step is MR since there are in total MR independent transmission paths. The maximum diversity of the second step is NR since there are in total NR independent transmission paths. Thus, the overall maximum diversity order is no higher than $\min\{M, N\}R$. When $M \neq N$, DSTC is optimal in diversity order. For the case of $M = N$, compared with the maximum diversity order, the diversity order degradation of DSTC is $R\log_e \log_e P/\log_e P$, which is negligible when P is very high. If the diversity order definition in (1.1) is used, since $\lim_{P\to\infty}(\log_e \log_e P/\log_e P) = 0$, it can be concluded that DSTC achieves full diversity order $\min\{M, N\}R$.

Till here, the basic DSTC, where each relay antenna linearly transforms its received signal only, has been introduced for multiple-antenna multiple-relay network. More general DSTC designs can be considered, similar to those in Sect. 2.2.4.

3.2.4 Code Design

The code design problem for DSTC in multiple-antenna relay network is to design both the space-time code at the transmitter **B** and the relay transformation matrices $\mathbf{A}_i$'s. Orthogonal design (OD) and quasi-orthogonal design (QOD) [1, 2, 5] can be applied similarly as in Sect. 2.3. In the following, we provide a few examples on the applications of real ODs in multiple-antenna relay networks.

The first example is on a network with one transmit antenna, two relay antennas, and two receive antennas, i.e., $M = 1, R = 2, N = 2$. We set $T = MR = 2$. The code at the transmitter, which is a vector for this network, is designed as

$$\mathbf{B} = \begin{bmatrix} b_1 & b_2 \end{bmatrix}^t,$$

where b_1 and b_2 are chosen as binary phase-shift keying (BPSK) signals, normalized according to (2.11). The matrices used at relays are designed as

$$\mathbf{A}_1 = \mathbf{I}_2 \quad \text{and} \quad \mathbf{A}_2 = \begin{bmatrix} 0 & -1 \\ 1 & 0 \end{bmatrix}.$$

The distributed space-time codeword formed at the receiver $\mathbf{S}$ is thus a 2×2 real OD [5].

The next example is on a network with $M = 2$, $R = 2$, $N = 1$, i.e., two transmit antennas, two relay antennas, and one receive antenna. We set $T = MR = 4$. The code at the transmitter, which is 4×2 for this network, is designed as

$$\mathbf{B} = \begin{bmatrix} b_1 & -b_2 \\ b_2 & b_1 \\ b_3 & -b_4 \\ b_4 & b_3 \end{bmatrix},$$

where b_1, b_2, b_3, b_4 are also BPSK signals, normalized according to (2.11). The matrices used at relays are designed as

$$\mathbf{A}_1 = \mathbf{I}_4 \quad \text{and} \quad \mathbf{A}_2 = \begin{bmatrix} 0 & 0 & -1 & 0 \\ 0 & 0 & 0 & 1 \\ 1 & 0 & 0 & 0 \\ 0 & 1 & 0 & 0 \end{bmatrix}.$$

The distributed space-time codeword formed at the receiver can be shown straightforwardly to be a 4×4 real OD [5].

The final example is on a network with $M = 2$, $R = 1$, $N = 2$, i.e., two transmit antenna, one relay antenna, and two receive antennas. We set $T = MR = 2$. The code at the transmitter is designed as

$$\mathbf{B} = \begin{bmatrix} b_1 & -b_2 \\ b_2 & b_1 \end{bmatrix},$$

where b_1 and b_2 are BPSK signals, normalized according to (2.11). The matrix used at the relay is set to be $\mathbf{I}_2$. The distributed space-time codeword formed at the receiver $\mathbf{S}$ is again a 2×2 real OD [5].

3.3 Combination of DSTC and MRC

In the DSTC scheme explained in Sect. 3.2, each relay antenna linearly processes the signal vector it receives independently. This is not optimal for networks whose relays have multiple co-located antennas. For antennas co-located at the same relay, their transmitted and received vectors can be processed jointly, which, conceivably, can lead to better performance. In this section, we use maximum-ratio combining (MRC) for the joint processing of the vectors received at co-located antennas before the linear processing designated for DSTC [3]. It is shown through both simulation and theoretical analysis that this combination of MRC and DSTC can improve the network reliability.

In what follows, we study the combination of MRC and DSTC for networks with single transmit antenna, single receive antenna, and a single multiple-antenna relay. Then extensions to multiple-antenna single-relay network and the general multiple-antenna multiple-relay network are considered.

3.3.1 MRC-DSTC for Network with Single Transmit Antenna, Single Receive Antenna, and Single Multiple-Antenna Relay

This subsection is devoted to networks with single antenna at the transmitter, single antenna at the receiver, and single relay equipped with R antennas. A diagram of this network is shown in Fig. 3.2. Denote the channel vector from the transmitter to the relay as $\mathbf{f}$, which is $1 \times R$, and the channel vector from the relay to the receiver as $\mathbf{g}$, which is $R \times 1$. The ith entries of $\mathbf{f}$ and $\mathbf{g}$ are the channels from the transmitter to relay antenna i and from relay antenna i to the receiver, respectively. There is no direct link. Assume that all channels follow Rayleigh flat-fading with the distribution $\mathcal{CN}(0, 1)$ and the block-fading model with coherence interval T. Assume also that the relay knows $\mathbf{f}$; the receiver knows $\mathbf{g}$ and $\|\mathbf{f}\|_F$. This CSI requirement can be fulfilled via training and channel estimation. Details can be found in [3].

MRC-DSTC Protocol

Information bits are encoded into a T-dimensional vector $\mathbf{b}$ with the normalization $\mathbb{E}(\mathbf{b}^*\mathbf{b}) = 1$. To send $\mathbf{b}$ from the transmitter to the receiver, two steps are needed. In the first step, the transmitter sends $\sqrt{P_s T}\mathbf{b}$. P_s is the average power per transmission of the transmitter. The signal vector received at the ith relay antenna is denoted as $\mathbf{r}_i$. The noise vector at the ith relay antenna is denoted as $\mathbf{n}_{r,i}$. Thus,

$$[\mathbf{r}_1 \ \cdots \ \mathbf{r}_R] = \sqrt{P_s T}\mathbf{b}\mathbf{f} + [\mathbf{n}_{r,1} \ \cdots \ \mathbf{n}_{r,R}]. \tag{3.25}$$

Fig. 3.2 Network with single transmit antenna, single receive antenna, and a single relay with R antennas

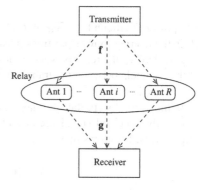

After receiving $\mathbf{r}_i$'s, the relay conducts signal processing to obtain $\mathbf{t}_i$'s. In the original DSTC, $\mathbf{t}_i$ is designed to be a linear function of $\mathbf{r}_i$ and its conjugate $\overline{\mathbf{r}_i}$. Thus, the transmitted and received signals of each relay antenna is processed independently regardless of the fact that the antennas are co-located on the same relay. In fact, $\mathbf{t}_i$ can be designed to be dependent on all $\mathbf{r}_1, \dots, \mathbf{r}_R$. Based on this idea, the following design that combines MRC and DSTC is proposed [3]. The relay first conducts the MRC of $\mathbf{r}_1, \dots, \mathbf{r}_R$ to get a less corrupted version of the transmitted vector, then conducts DSTC to linearly transform the combined signal vector onto different subspaces. In the following, this scheme is addressed as MRC-DSTC.

Here are the details of the relay processing. For the MRC, the relay, who knows $\mathbf{f}$, right-multiplies $\begin{bmatrix} \mathbf{r}_1 & \cdots & \mathbf{r}_R \end{bmatrix}$ with $\mathbf{f}^*/\|\mathbf{f}\|_F$ to obtain $\mathbf{r}$. From (3.25), we have

$$\mathbf{r} \triangleq \begin{bmatrix} \mathbf{r}_1 & \cdots & \mathbf{r}_R \end{bmatrix} \frac{\mathbf{f}^*}{\|\mathbf{f}\|_F} = \sqrt{P_s T}\|\mathbf{f}\|_F \mathbf{b} + \mathbf{n}_r, \tag{3.26}$$

where $\mathbf{n}_r \triangleq \begin{bmatrix} \mathbf{n}_{r,1} & \cdots & \mathbf{n}_{r,R} \end{bmatrix} \frac{\mathbf{f}^*}{\|\mathbf{f}\|_F}$. For any given realization of $\mathbf{f}$, since entries of $\mathbf{n}_{r,i}$'s are i.i.d. $\mathcal{CN}(0, 1)$, it can be shown straightforwardly that entries of $\mathbf{n}_r$ are also i.i.d. $\mathcal{CN}(0, 1)$. The MRC process mitigates the relay noise propagation. This can be seen by calculating the signal-to-noise-ratio (SNR) values. It can be shown straightforwardly that the SNR of the combined signal $\mathbf{r}$, averaged over channel statistics, is increased by R-fold compared to each $\mathbf{r}_i$.

After MRC, DSTC is conducted using the combined signal. Define

$$\alpha_{\text{MRC}} \triangleq \frac{P_r}{1 + P_s \|\mathbf{f}\|_F^2}. \tag{3.27}$$

The transmitted vector of each relay antenna, $\mathbf{t}_i$, is designed to be a linear function of $\mathbf{r}$ and $\overline{\mathbf{r}}$ as:

$$\mathbf{t}_i = \sqrt{\alpha_{\text{MRC}}}(\mathbf{A}_i \mathbf{r} + \mathbf{B}_i \overline{\mathbf{r}}), \tag{3.28}$$

where $\mathbf{A}_i$ and $\mathbf{B}_i$ are $T \times T$ matrices with the following normalization

$$\text{tr}(\mathbf{A}_i \mathbf{A}_i^* + \mathbf{B}_i \mathbf{B}_i^*) = T. \tag{3.29}$$

The power coefficient $\sqrt{\alpha_{\text{MRC}}}$ in (3.27) and the normalization in (3.29) ensure that P_r is the average power per transmission of every relay antenna. Compared to the original DSTC, the relay power coefficient for MRC-DSTC is adaptive to the quality of the channels between the transmitter and the relay through $\|\mathbf{f}\|_F$, as can be seen from in (3.27).

In the second step, the ith relay antenna sends $\mathbf{t}_i$ to the receiver simultaneously. Denote the received signal vector and the noise vector at the receiver as $\mathbf{x}$ and $\mathbf{n}_d$ respectively. It follows that

$$\mathbf{x} = [\mathbf{t}_1 \ \cdots \ \mathbf{t}_R]\mathbf{g} + \mathbf{n}_d. \tag{3.30}$$

Define

$$\beta_{\text{MRC}} \triangleq \frac{P_s P_r T}{1 + P_s \|\mathbf{f}\|_F^2}$$

By using (3.26) and (3.28) in (3.30), we have

$$\mathbf{x} = \sqrt{\beta_{\text{MRC}}} \|\mathbf{f}\|_F \mathbf{Sg} + \mathbf{w}, \tag{3.31}$$

where

$$\mathbf{S} \triangleq \left[\mathbf{A}_1 \mathbf{b} + \mathbf{B}_1 \overline{\mathbf{b}} \quad \cdots \quad \mathbf{A}_R \mathbf{b} + \mathbf{B}_R \overline{\mathbf{b}} \right] \tag{3.32}$$

and

$$\mathbf{w} \triangleq \sqrt{\alpha_{\text{MRC}}} \left[\mathbf{A}_1 \mathbf{n}_r + \mathbf{B}_1 \overline{\mathbf{n}_r} \quad \cdots \quad \mathbf{A}_R \mathbf{n}_r + \mathbf{B}_R \overline{\mathbf{n}_r} \right] \mathbf{g} + \mathbf{n}_d.$$

$\mathbf{S}$ is the $T \times R$ space-time codeword. $\mathbf{w}$ is the equivalent noise vector. When the noises at the relays and the receiver are assumed to be i.i.d. following $\mathcal{CN}(0, 1)$, $\mathbf{w}$ can be proved to be circularly symmetric Gaussian with zero-mean. Its covariance matrix can be calculated to be

$$\mathbf{R_w} \triangleq \mathbf{I}_T + \alpha_{\text{MRC}} \left(\left[\mathbf{A}_1 \cdots \mathbf{A}_R \right] (\mathbf{gg}^* \otimes \mathbf{I}_T) \left[\mathbf{A}_1 \cdots \mathbf{A}_R \right]^* \right.$$
$$\left. + \left[\mathbf{B}_1 \cdots \mathbf{B}_R \right] (\mathbf{gg}^* \otimes \mathbf{I}_T) \left[\mathbf{B}_1 \cdots \mathbf{B}_R \right]^* \right). \tag{3.33}$$

It should be clarified that the space-time codeword for MRC-DSTC given in (3.32) has essential difference to that for DSTC given in (2.17). The code matrix in (3.32) is $T \times R$, while the code in (2.17) is $T \times MR$. With slight abuse of notation but for the consistency of the presentation, the same notation $\mathbf{S}$ is used. The same goes to the notation for the equivalent noise $\mathbf{w}$ and its covariance matrix.

As the receiver knows $\|\mathbf{f}\|_F$ and $\mathbf{g}$, the ML decoding of MRC-DSTC can be derived straightforwardly to be

$$\arg \min_{\mathbf{b}} \left(\mathbf{x} - \sqrt{\beta_{\text{MRC}}} \|\mathbf{f}\|_F \mathbf{Sg} \right)^* \mathbf{R_w}^{-1} \left(\mathbf{x} - \sqrt{\beta_{\text{MRC}}} \|\mathbf{f}\|_F \mathbf{Sg} \right).$$

Diversity Order Analysis

Now we analyze the PEP of MRC-DSTC, based on which the diversity order can be observed. Due to the similarities between MRC-DSTC and DSTC, derivations of the PEP for DSTC in Sect. 2.2.3 can be used immediately for MRC-DSTC to obtain an upper bound on the average PEP of mistaking the information vector $\mathbf{b}_k$ by another information vector $\mathbf{b}_l$ as:

$$\mathbb{P} (\mathbf{b}_k \rightarrow \mathbf{b}_l) \leq \mathbb{E}_{\mathbf{f}, \mathbf{g}} e^{-\frac{\beta_{\text{MRC}}}{4} \|\mathbf{f}\|_F^2 \mathbf{g}^* (\mathbf{S}_k - \mathbf{S}_l)^* \mathbf{R_w}^{-1} (\mathbf{S}_k - \mathbf{S}_l) \mathbf{g}}, \tag{3.34}$$

where $\mathbf{S}_k$ and $\mathbf{S}_l$ are the two space-time codewords corresponding to $\mathbf{b}_k$ and $\mathbf{b}_l$. By noticing that $\mathbf{gg}^* \preceq \|\mathbf{g}\|_F^2 \mathbf{I}_R$, it can be shown from (3.33) that

$$\mathbf{R_w} \preceq \mathbf{I}_T + \alpha_{\mathrm{MRC}} \|\mathbf{g}\|_F^2 \sum_{i=1}^{R} \left(\mathbf{A}_i \mathbf{A}_i^* + \mathbf{B}_i \mathbf{B}_i^*\right) \preceq (1 + \alpha_{\mathrm{MRC}} RT \|\mathbf{g}\|_F^2) \mathbf{I}_T.$$

Denote the minimum singular value of $\mathbf{M}_{kl}$ as $\sigma_{\min}^2$. When $T \geq R$ and the space-time code is fully diversity, we have $\sigma_{\min}^2 > 0$. Let $\xi \triangleq \min\{\|\mathbf{f}\|_F^2, \|\mathbf{g}\|_F^2\}$. Thus,

$$\frac{\|\mathbf{f}\|_F^2 \|\mathbf{g}\|_F^2}{1 + P_s \|\mathbf{f}\|_F^2 + R P_r T \|\mathbf{g}\|_F^2} \geq \frac{\xi^2}{1 + (P_s + R P_r T)\xi}.$$

From (3.34), the PEP can be further upper bounded as

$$\mathbb{P}(\mathbf{b}_k \to \mathbf{b}_l) \leq \mathbb{E}_{\mathbf{f},\mathbf{g}} e^{-\frac{\beta_{\mathrm{MRC}}}{4(1+\alpha_{\mathrm{MRC}} RT \|\mathbf{g}\|_F^2)} \|\mathbf{f}\|_F^2 \mathbf{g}^* \mathbf{M}_{kl} \mathbf{g}}$$

$$\leq \mathbb{E}_{\mathbf{f},\mathbf{g}} e^{-\frac{P_s P_r T}{4(1+P_s \|\mathbf{f}\|_F^2 + R P_r T \|\mathbf{g}\|_F^2)} \|\mathbf{f}\|_F^2 \mathbf{g}^* \mathbf{M}_{kl} \mathbf{g}}$$

$$\leq \mathbb{E}_{\mathbf{f},\mathbf{g}} e^{-\frac{P_s P_r T \sigma_{\min}^2}{4(1+P_s \|\mathbf{f}\|_F^2 + R P_r T \|\mathbf{g}\|_F^2)} \|\mathbf{f}\|_F^2 \|\mathbf{g}\|_F^2}$$

$$\leq \mathbb{E}_{\xi} e^{-\frac{P_s P_r T \sigma_{\min}^2 \xi^2}{4[1+(P_s + R P_r T)\xi]}}. \tag{3.35}$$

Both $\|\mathbf{f}\|_F^2$ and $\|\mathbf{g}\|_F^2$ have Gamma distribution with degree R. Denote their PDF as $p(x)$ and cumulative distribution function (CDF) as $P(x)$. Thus, $p(x) = \frac{x^{R-1}}{\Gamma(R)} e^{-x}$. The PDF of ξ, denoted as $q(x)$, can thus be proved to satisfy $q(x) = 2p(x)(1 - P(x)) < 2p(x)$. Using this in (3.35), we have

$$\mathbb{P}(\mathbf{b}_k \to \mathbf{b}_l) \leq \int_0^{\infty} e^{-\frac{P_s P_r T \sigma_{\min}^2 x^2}{4[1+(P_s + R P_r T)x]}} q(x) dx$$

$$\leq 2 \int_0^{\infty} e^{-\frac{P_s P_r T \sigma_{\min}^2 x^2}{4[1+(P_s + R P_r T)x]}} p(x) dx$$

$$\leq 2 \int_{\frac{1}{P_s + R P_r T}}^{\infty} e^{-\frac{P_s P_r T \sigma_{\min}^2}{8(P_s + R P_r T)} x} p(x) dx + 2 \int_0^{\frac{1}{P_s + R P_r T}} p(x) dx$$

$$\leq \frac{2}{\Gamma(R)} \int_0^{\infty} e^{-\frac{P_s P_r T \sigma_{\min}^2}{8(P_s + R P_r T)} x} x^{R-1} dx + \frac{2}{\Gamma(R)} \int_0^{\frac{1}{P_s + R P_r T}} x^{R-1} dx$$

$$= \frac{2}{\left[\frac{P_s P_r T \sigma_{\min}^2}{8(P_s + R P_r T)}\right]^R} + \frac{2}{\Gamma(R+1)} \frac{1}{(P_s + R P_r T)^R}.$$

When both P_s and P_r scale as P, i.e., $P_s = \mathcal{O}(P)$ and $P_r = \mathcal{O}(P)$, the above PEP upper bound scales as P^{-R}. Therefore, there exists a constant c such that

$\mathbb{P}(\mathbf{b}_k \rightarrow \mathbf{b}_l) \leq cP^{-R}$. By using the diversity definitions in (1.1) and (1.2), we can conclude that MRC-DSTC achieves full diversity order, which is R. This result is written in the following theorem.

Theorem 3.2. *Assume that $T \geq R$ and the distributed space-time code is fully diversity. For networks with single transmit antenna, single receive antenna, and one single relay with R antennas, the diversity order of the MRC-DSTC scheme is R.*

Comparison of MRC-DSTC and DSTC

In this part, we compare MRC-DSTC with DSTC. While the MRC in MRC-DSTC can only be conducted for relay antennas that are co-located at the same relay, DSTC applies for networks with distributed relay antennas. To compare the performance, consider two relay networks with single transmit antenna, single receive antenna, and R relay antennas. But for Network 1, the R relay antennas are co-located at the same relay and MRC-DSTC is used; while for Network 2, the R relay antennas can be distributively located at R relays and DSTC is used. It is shown in Chap. 2 that an upper bound on the PEP of Network 2 under DSTC scales as $P^{-R} \log_e^R P$ and the diversity order is $R(1 - \log_e \log_e P / \log_e P)$. For Network 1 under MRC-DSTC, the $\log_e^R P$ factor in the PEP upper bound is eliminated and the diversity order is R. One account of this diversity order improvement is that MRC at the multiple-antenna relay suppresses the relay noise.

For the special case of $R = 1$, there is no combining gain in MRC-DSTC; but the PEP of MRC-DSTC scales as P^{-1}, while that of DSTC scales as $P^{-1} \log_e P$. The difference is in the relay power coefficient. For MRC-DSTC, the relay power coefficient $\sqrt{\alpha_{\text{MRC}}}$ is a function of $\|\mathbf{f}\|_F$, which represents the quality of the first transmission step. For DSTC, the constant channel-independent power coefficient $\sqrt{\alpha}$ is used. Although the two result in the same average relay power, the former is adaptive to the channel quality of the first step, thus has better performance.

It should be clarified that to achieve this performance improvement, MRC-DSTC requires partial CSI at the relay and cross-talk among relay antennas.

3.3.2 MRC-DSTC for Multiple-Antenna Single-Relay Network

Consider a multiple-antenna single-relay network with M transmit antennas, N receive antennas, and R relay antennas co-located at one relay node. A diagram is shown in Fig. 3.3. Denote the $M \times 1$ channel vector from the transmitter and the ith antenna of the relay as $\mathbf{f}_i$. Denote the $M \times R$ transmitter-relay channel matrix as $\mathbf{F}$. Its ith column is $\mathbf{f}_i$. Denote the $1 \times N$ channel vector from the ith antenna of the relay to the receiver as $\mathbf{g}_i$. Denote the $R \times N$ relay-receiver channel matrix as $\mathbf{G}$. Its ith row is $\mathbf{g}_i$. To use MRC-DSTC, a combining scheme for the relay is needed. When $M = 1$, the MRC of the signal vectors the relay antennas receive can be performed straightforwardly as shown in Sect. 3.3.1. But when there are multiple transmit antennas, the MRC is less straightforward.

Fig. 3.3 Multiple-Antenna single-relay network

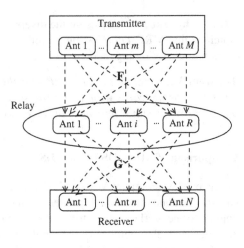

When $M > 1$, the first transmission step can be seen as a multiple-antenna system, and the transmitter should send a $T \times M$ space-time code-matrix $\mathbf{B}$. However, fully diverse receiver combining schemes for space-time coded communications are unavailable.

To overcome this difficulty, the special case of orthogonal space-time codes that allow decoupled symbol-wise decoding is considered. Their special structure leads to convenient combining at the relay. For presentation simplicity, here, we only consider orthogonal codes whose symbol-rates are 1. Such codes include Alamouti design, real ODs, and their straightforward extensions [2, 5].

The details of the two-step MRC-DSTC transmission protocol are as follows. Let $b_1, \cdots, b_T$ be the information symbols and $\mathbf{b} \triangleq [b_1 \ \cdots \ b_T]^t$ is the T-dimensional information vector. First, $\mathbf{b}$ is encoded into a $T \times M$ orthogonal space-time codeword $\mathbf{B}$. In Step 1, the transmitter sends $\sqrt{P_s T/M}\mathbf{B}$ with P_s the power of the transmitter. Relay antenna i gets

$$\mathbf{r}_i = \sqrt{\frac{P_s T}{M}}\mathbf{B}\mathbf{f}_i + \mathbf{n}_{r,i}.$$

Since the code-matrix $\mathbf{B}$ is orthogonal and has symbol-wise decoding, the relay can process $\mathbf{r}_i$ using the OD structure to obtain $\check{\mathbf{r}}_i$, which satisfies the following equation:

$$\check{\mathbf{r}}_i = \sqrt{\frac{P_s T}{M}}\|\mathbf{f}_i\|_F \mathbf{b} + \check{\mathbf{n}}_{r,i}, \tag{3.36}$$

where $\check{\mathbf{n}}_{r,i}$ follows $\mathcal{CN}(\mathbf{0}_T, \mathbf{I}_T)$. For more details on this step of processing and an illustrative example, please see [3]. From (3.36), we have

$$\left[\check{\mathbf{r}}_1 \ \cdots \ \check{\mathbf{r}}_R\right] = \sqrt{\frac{P_s T}{M}}\mathbf{b}\left[\|\mathbf{f}_1\|_F \ \cdots \ \|\mathbf{f}_R\|_F\right] + \left[\check{\mathbf{n}}_{r,1} \ \cdots \ \check{\mathbf{n}}_{r,R}\right]. \tag{3.37}$$

The relay then conducts the MRC of $\check{\mathbf{r}}_i$'s, where $\begin{bmatrix} \check{\mathbf{r}}_1 & \cdots & \check{\mathbf{r}}_R \end{bmatrix}$ is right-multiplied with $\frac{1}{\|\mathbf{F}\|_F} \begin{bmatrix} \|\mathbf{f}_1\|_F & \cdots & \|\mathbf{f}_R\|_F \end{bmatrix}^t$ to obtain the combined signal vector $\mathbf{r}$. From (3.37), we have

$$\mathbf{r} \triangleq \frac{1}{\|\mathbf{F}\|_F} \sum_{i=1}^{R} \|\mathbf{f}_i\|_F \hat{\mathbf{r}}_i = \sqrt{\frac{P_s T}{M}} \|\mathbf{F}\|_F \mathbf{b} + \frac{1}{\|\mathbf{F}\|_F} \sum_{i=1}^{R} \|\mathbf{f}_i\|_F \check{\mathbf{n}}_{r,i}. \qquad (3.38)$$

It can be proved that the noise term $\frac{1}{\|\mathbf{F}\|_F} \sum_{i=1}^{R} \|\mathbf{f}_i\|_F \check{\mathbf{n}}_{r,i}$ in (3.38) still follows $\mathcal{CN}(\mathbf{0}_T, \mathbf{I}_T)$.

After obtaining the combined vector $\mathbf{r}$. Antenna i of the relay linearly processes $\mathbf{r}$ and $\bar{\mathbf{r}}$ to obtain $\mathbf{t}_i$ as:

$$\mathbf{t}_i = \sqrt{\frac{P_r}{1 + \|\mathbf{F}\|_F^2 P_s/M}} \left(\mathbf{A}_i \mathbf{r} + \mathbf{B}_i \bar{\mathbf{r}} \right), \qquad (3.39)$$

where $\mathbf{A}_i$ and $\mathbf{B}_i$ are $T \times T$ pre-determined relay transformation matrices satisfying the normalization in (3.29).

In Step 2, Antenna i of the relay sends $\mathbf{t}_i$. All relay antennas send information simultaneously. Denote the matrix of the received signals and the matrix of the noises at the receiver as $\mathbf{X}$ and $\mathbf{N}_d$ respectively. It follows that

$$\mathbf{X} = [\mathbf{t}_1 \cdots \mathbf{t}_R]\mathbf{G} + \mathbf{N}_d. \qquad (3.40)$$

With slight abuse of notation but for the consistency of the presentation, define

$$\alpha_{\mathrm{MRC}} \triangleq \sqrt{\frac{P_r}{1 + \|\mathbf{F}\|_F^2 P_s/M}}, \quad \beta_{\mathrm{MRC}} \triangleq \sqrt{\frac{P_s P_r T}{M + \|\mathbf{F}\|_F^2 P_s}}.$$

Using (3.38) and (3.39) in (3.40), we can write the transceiver equation as

$$\mathbf{X} = \sqrt{\beta_{\mathrm{MRC}}} \|\mathbf{F}\|_F \mathbf{S}\mathbf{G} + \mathbf{W}, \qquad (3.41)$$

where $\mathbf{S} \triangleq \begin{bmatrix} \mathbf{A}_1\mathbf{b} + \mathbf{B}_1\bar{\mathbf{b}} & \cdots & \mathbf{A}_R\mathbf{b} + \mathbf{B}_R\bar{\mathbf{b}} \end{bmatrix}$ and

$$\mathbf{W} \triangleq \sqrt{\alpha_{\mathrm{MRC}}} \begin{bmatrix} \mathbf{A}_1\mathbf{v} + \mathbf{B}_1\bar{\mathbf{v}} & \cdots & \mathbf{A}_R\mathbf{v} + \mathbf{B}_R\bar{\mathbf{v}} \end{bmatrix}\mathbf{G} + \mathbf{N}_d.$$

$\mathbf{S}$, which is $T \times R$, is the space-time codeword.

To obtain the ML decoding and thus conduct the PEP analysis, columns of the system equation in (3.41) are stacked into a column vector to get the vector equation:

$$\mathrm{vec}(\mathbf{X}) = \sqrt{\beta_{\mathrm{MRC}}}(\mathbf{I}_N \otimes \mathbf{S}) \|\mathbf{F}\|_F \mathrm{vec}(\mathbf{G}) + \mathrm{vec}(\mathbf{W}).$$

Define $\mathbf{h} \triangleq \|\mathbf{F}\|_F \text{vec}(\mathbf{G})$. The ML decoding is thus

$$\arg \min_{\mathbf{b}} \left[\text{vec}(\mathbf{X}) - \sqrt{\beta_{\text{MRC}}}(\mathbf{I}_N \otimes \mathbf{S})\mathbf{h} \right]^* \mathbf{R}^{-1}_{\text{vec}(\mathbf{W})} \left[\text{vec}(\mathbf{X}) - \sqrt{\beta_{\text{MRC}}}(\mathbf{I}_N \otimes \mathbf{S})\mathbf{h} \right],$$

$$(3.42)$$

where $\mathbf{R}_{\text{vec}(\mathbf{W})}$ is the covariance matrix of $\text{vec}(\mathbf{W})$. If $\mathbf{R}_{\text{vec}(\mathbf{W})}$ is decomposed into $T \times T$ blocks, its (i, j)-th block is

$$\delta_{ij}\mathbf{I}_T + \alpha_{\text{MRC}} \sum_{k,l=1}^{R} g_{ki}\overline{g_{lj}}(\mathbf{A}_k\mathbf{A}_l^* + \mathbf{B}_k\mathbf{B}_l^*),$$

where g_{ij} is the (i, j)th entry of $\mathbf{G}$.

With the ML decoding in (3.42), the PEP and diversity order of MRC-DSTC can be analyzed. The result is stated in the following Theorem.

Theorem 3.3. *Assume that $T \geq R$ and the space-time code is fully diverse. For a network with M transmit antennas, N receive antennas, and one R-antenna relay, MRC-DSTC achieves diversity order $\min\{M, N\}R$.*

The proof of this theorem is similar to that of Theorem 3.2 in Sect. 3.3.1. The details can be found in [3].

Comparing the diversity order result in Theorem 3.2 with that of DSTC in Chap. 3, we can see that for the case of $M = N$ (the same number of transmit and receive antennas), MRC-DSTC eliminates the diversity order degradation $R \log_e \log_e P / \log_e P$, and achieves full diversity $\min\{M, N\}R$. The improvement is due to both the combining of the signals received at different relay antennas and the channel-dependent power coefficient at the relay. For the case of $M \neq N$, MRC-DSTC and DSTC have the same full diversity but the former has extra combining gain, thus better performance.

3.3.3 MRC-DSTC for a General Multiple-Antenna Multiple-Relay Network

The generalization of MRC-DSTC to the most general multiple-antenna multiple-relay network was studied in [3], where the protocol is similar to that of the multiple-antenna single-relay network. However, a channel-independent (fixed-gain) relay power coefficient is used for the tractability of the diversity order analysis. The first and second steps of transmissions are the same as before. For the relay processing, after Step 1, Relay k conducts MRC of the received signals of all its antennas to obtain $\mathbf{r}_k$. The transmitted signal of the ith antenna of Relay k for the second step, $\mathbf{t}_{k,i}$, is then design as

$$t_{k,i} = \sqrt{\frac{P_r}{1 + RP_s}} \left(\mathbf{A}_{k,i} \mathbf{r}_k + \mathbf{B}_{k,i} \bar{\mathbf{r}}_k \right), \tag{3.43}$$

where $\mathbf{A}_{k,i}$ and $\mathbf{B}_{k,i}$ are the transformation matrices at the ith antenna of Relay k. The relay power coefficient $\sqrt{\frac{P_r}{1+RP_s}}$ in (3.43) is a constant, independent of the channel quality. The following diversity order result of MRC-DSTC was proved [3].

Theorem 3.4. *Assume that $T \geq R$ and the space-time code is fully diverse. For a network with M transmit antennas, N receive antennas, a total of R relay antennas located on K relay nodes, an achievable diversity order of MRC-DSTC with fixed-gain relay power coefficient is*

$$d_{\text{MRC-DSTC}} = \begin{cases} \min\{M, N\}R & \text{if } M \neq N \\ MR - K \frac{\log_e \log_e P}{\log_e P} & \text{if } M = N \end{cases}.$$

The proof of this theorem can be found in [3].

When $M = N$, the diversity order of DSTC is $MR - R \frac{\log_e \log_e P}{\log_e P}$. Comparing with the result in Theorem 3.4, we can see that the diversity order of MRC-DSTC is improved by $(R - K) \frac{\log_e \log_e P}{\log_e P}$, thanks to the MRC conducted at the relays. The diversity order of MRC-DSTC with channel-dependent (variable gain) relay power coefficient has not been derived.

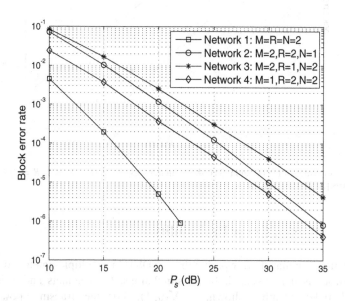

Fig. 3.4 Performance of DSTC in 4 networks: *(1)* $M = R = N = 2$; *(2)* $M = 2$, $R = 2$, $N = 1$; *(3)* $M = 2$, $R = 1$, $N = 2$; and *(4)* $M = 1$, $R = 2$, $N = 2$. ODs with BPSK modulation are used

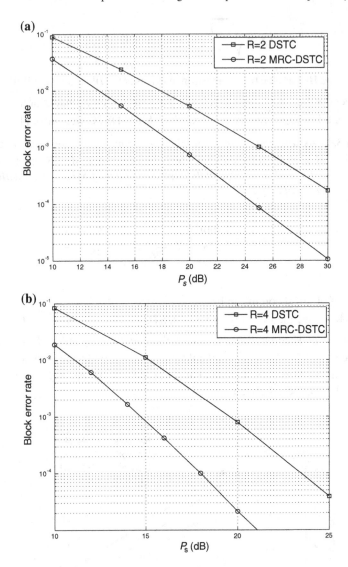

Fig. 3.5 Performance of DSTC and DSTC-MRC in networks with $M = N = 1$, $R = 2, 4$, random code and BPSK. (a) $R = 2$ (b) $R = 4$

3.4 Simulation on Error Probability

In this section, simulated error probability of DSTC in multiple-antenna multiple-relay network is demonstrated. In all simulations, the channels and noises are generated independently following $\mathcal{CN}(0, 1)$. For the transmit power, we set $P_s = R P_r$.

Figure 3.4 shows the block error rates of 4 networks: (1) $M = R = N = 2$; (2) $M = 2, R = 2, N = 1$; (3) $M = 2, R = 1, N = 2$; and (4) $M = 1, R = 2, N = 2$. Real ODs are used for the distributed space-time code. That is, for Network 1 and Network 2, T is 4 and the 4×4 real OD is used; for Network 3 and Network 4, T is 2 and the 2×2 real OD is used. Note that if we set $T = 4$ for Network 3 and Network 4 and use 4×2 real OD, the same performance will be obtained. BPSK is used to modulate the information symbols. So the overall transmission rate is 1/2 bit per transmission. The figure shows that when P_s is large, the diversity order of Network 2 and Network 4 is 2. The diversity order of Network 3 is slightly less than 2 but approaches 2 as P_s increases. For Network 1, the diversity order approaches 4 as P_s increases.

In Fig. 3.5, block error rates of two networks with $M = N = 1$, and $R = 2, 4$ are shown. For MRC-DSTC, the R relay antennas are assumed to be co-located at the same relay so MRC is possible; while for DSTC, the relay antennas can be distributively located. Random code explained in Sect. 2.4 is used to avoid the effect of code design and focus on the diversity order only. It can be seen that the block error rate curves for MRC-DSTC are slightly steeper than the corresponding DSTC curves, indicating larger diversity orders. This is consistent with the diversity order derivation in Sect. 3.3. In addition to the improvement in diversity order, it can also be seen from the plots that MRC at relays significant improves the network performance. At the block error rate level of 2×10^{-4}, MRC-DSTC is approximately 6dB better for both networks.

References

1. Alamouti SM (1998) A simple transmitter diversity scheme for wireless communications. IEEE J on Selected Areas in Communications, 16:1451–1458.
2. Jafarkhani H (2005) Space-Time Coding: Theory and Practice. Cambridge Academic Press.
3. Jing Y (2010) Combination of MRC and distributed space-time coding in networks with multiple-antenna relays. IEEE T on Wireless Communications, 9:2550–2559.
4. Jing Y and Hassibi B (2008) Cooperative diversity in wireless relay networks with multiple-antenna nodes. EURASIP J on Advanced, Signal Process, 19 pages doi:10.1155/2008/254573
5. Tarokh V, Jafarkhani H, and Calderbank AR (1999) Space-time block codes from orthogonal designs. IEEE T on Information Theory, 45:1456–1467.

Chapter 4
Differential Distributed Space-Time Coding

Abstract This chapter is on the differential use of distributed space-time coding (DSTC). First, the transmission protocol and decoding of differential DSTC are explained in Sect. 4.1. Then several code designs are introduced in Sect. 4.2. Finally, simulated error probability of differential DSTC is shown in Sect. 4.3.

4.1 Transmission Protocol and Decoding

Differential distributed space-time coding (DSTC) [5, 6, 8] is the differential use of DSTC. It can also be seen as the generalization of differential space-time coding (reviewed in Sect. 2.1) to relay networks. The presentation and materials in this chapter follow the work in [5].

Consider the single-antenna multiple-relay network shown in Fig. 2.2 in Chap. 2. We use the DSTC protocol described in Sect. 2.2.4, where the transmitted signal vector of a relay antenna is either a linear transformation of its received signal vector or the conjugate of its received signal vector, i.e., either $\mathbf{A}_i = 0, \mathbf{B}_i$ is unitary or $\mathbf{B}_i = 0, \mathbf{A}_i$ is unitary. Let $T_1 = T_2 = T = R$ and call the transmission of T symbols a block. Therefore, a block contains $2T$ time slots with T time slots for each step. The differential DSTC scheme uses two DSTC blocks that overlap by one block, where each block acts as a reference for the next.

The details of the differential DSTC protocol are as follows. Denote the $T \times 1$ information vector sent in the τth block as $\mathbf{b}_{(\tau)}$. We start with the first block, then elaborate the transmission of the τth block for generality. For the first block, a reference vector that satisfies $\mathbb{E}\{\mathbf{b}_{(0)}^* \mathbf{b}_{(0)} = 1\}$ is sent using DSTC, for example, $\mathbf{b}_{(0)} = \begin{bmatrix} 1 \ 0 \cdots 0 \end{bmatrix}^t$. For the $(\tau-1)$th block, $\mathbf{b}_{(\tau-1)}$ is sent using DSTC. Define

$$\begin{cases} \check{\mathbf{A}}_i = \mathbf{A}_i, \ \mathbf{b}_{(\tau-1)}^{(i)} = \mathbf{b}_{(\tau-1)} & \text{if } \mathbf{B}_i = 0 \\ \check{\mathbf{A}}_i = \mathbf{B}_i, \ \mathbf{b}_{(\tau-1)}^{(i)} = \overline{\mathbf{b}_{(\tau-1)}} & \text{if } \mathbf{A}_i = 0 \end{cases}.$$

Y. Jing, *Distributed Space-Time Coding*, SpringerBriefs in Computer Science,   69
DOI: 10.1007/978-1-4614-6831-8_4, © The Author(s) 2013

From (2.46)–(2.49), the transceiver equation of the $(\tau - 1)$th block is

$$\mathbf{x}_{(\tau-1)} = \sqrt{\beta} \left[\check{\mathbf{A}}_1 \mathbf{b}_{(\tau-1)}^{(1)} \quad \cdots \quad \check{\mathbf{A}}_R \mathbf{b}_{(\tau-1)}^{(R)} \right] \mathbf{h}_{(\tau-1)} + \mathbf{w}_{(\tau-1)}, \qquad (4.1)$$

where $\mathbf{x}_{(\tau-1)}$ is the received signal vector, $\mathbf{w}_{(\tau-1)}$ is the equivalent noise vector, $\mathbf{h}_{(\tau-1)}$ is the equivalent end-to-end channel vector of block $\tau - 1$, and β is defined in (2.16).

Information is encoded into a set of $T \times T$ unitary matrices $\mathcal{U}$. For the τth block, a matrix $\mathbf{U}_{(\tau)}$ is selected from the set of code-matrices $\mathcal{U}$ based on the information to be sent. To communicate message $\mathbf{U}_{(\tau)}$ to the receiver, the signal vector sent by the transmitter is encoded differentially as

$$\mathbf{b}_{(\tau)} = \mathbf{U}_{(\tau)} \mathbf{b}_{(\tau-1)}. \qquad (4.2)$$

Note that having $\mathbf{U}_{(\tau)}$ unitary preserves the transmit power, because the Frobenius norm of a vector keeps unchanged after unitary transformation. By following the DSTC protocol, the transceiver equation of the τ'th block can be written as

$$\mathbf{x}_{(\tau)} = \sqrt{\beta} \left[\check{\mathbf{A}}_1 \check{\mathbf{b}}_{(\tau)}^{(1)} \quad \cdots \quad \check{\mathbf{A}}_R \check{\mathbf{b}}_{(\tau)}^{(R)} \right] \mathbf{h}_{(\tau)} + \mathbf{w}_{(\tau)}.$$

By using (4.2), it can be obtained that

$$\mathbf{x}_{(\tau)} = \sqrt{\beta} \left[\check{\mathbf{A}}_1 \check{\mathbf{U}}_{(\tau)} \check{\mathbf{b}}_{(\tau-1)}^{(1)} \quad \cdots \quad \check{\mathbf{A}}_R \check{\mathbf{U}}_{(\tau)} \check{\mathbf{b}}_{(\tau-1)}^{(R)} \right] \mathbf{h}_{(\tau)} + \mathbf{w}_{(\tau)},$$

where

$$\begin{cases} \check{\mathbf{U}}_{(\tau)} = \mathbf{U}_{(\tau)} & \text{if } \mathbf{B}_i = 0 \\ \check{\mathbf{U}}_{(\tau)} = \overline{\mathbf{U}_{(\tau)}} & \text{if } \mathbf{A}_i = 0 \end{cases}$$

If $\mathbf{U}_{(\tau)} \check{\mathbf{A}}_i = \check{\mathbf{A}}_i \check{\mathbf{U}}_{(\tau)}$, or equivalently,

$$\begin{cases} \mathbf{U}_{(\tau)} \mathbf{A}_i = \mathbf{A}_i \mathbf{U}_{(\tau)} \\ \mathbf{U}_{(\tau)} \mathbf{B}_i = \mathbf{B}_i \overline{\mathbf{U}_{(\tau)}} \end{cases}, \qquad (4.3)$$

it follows that

$$\mathbf{x}_{(\tau)} = \sqrt{\beta} \mathbf{U}_{(\tau)} \left[\check{\mathbf{A}}_1 \mathbf{b}_{(\tau-1)}^{(1)} \quad \cdots \quad \check{\mathbf{A}}_R \mathbf{b}_{(\tau-1)}^{(R)} \right] \mathbf{h}_{(\tau)} + \mathbf{w}_{(\tau)}. \qquad (4.4)$$

If the channels $\mathbf{f}$ and $\mathbf{g}$ keep constant for two blocks, i.e., $\mathbf{h}_{(\tau)} = \mathbf{h}_{(\tau-1)}$, from (4.1) and (4.4), we have

$$\mathbf{x}_{(\tau)} = \mathbf{U}_{(\tau)} \mathbf{x}_{(\tau-1)} + \tilde{\mathbf{w}}_{(\tau)},$$

where

$$\tilde{\mathbf{w}}_{(\tau)} = \mathbf{w}_{(\tau)} - \mathbf{U}_{(\tau)} \mathbf{w}_{(\tau-1)}.$$

Given an arbitrary realization of $\mathbf{g}$, it is shown in Sect. 2.2.2 that $\mathbf{w}_{(\tau-1)}$ and $\mathbf{w}_{(\tau)}$ are independent complex Gaussian random vectors with mean $\mathbf{0}$ and covariance matrix $\left(1 + \alpha \|\mathbf{g}\|_F^2\right) \mathbf{I}_T$. Thus, $\tilde{\mathbf{w}}_{(\tau)}$ is a Gaussian random vector whose mean is $\mathbf{0}$ and whose covariance matrix is $2\left(1 + \alpha \|\mathbf{g}\|_F^2\right) \mathbf{I}_T$. The ML decoding of differential DSTC is thus the following:

$$\hat{\mathbf{U}}_{(\tau)} = \arg\max_{\mathbf{U}_{(\tau)}} \left[2\left(1 + \alpha \|\mathbf{g}\|_F^2\right)\right]^{-1} \left\|\mathbf{x}_{(\tau)} - \mathbf{U}_{(\tau)}\mathbf{x}_{(\tau-1)}\right\|_F^2$$

$$= \arg\max_{\mathbf{U}_{(\tau)}} \left\|\mathbf{x}_{(\tau)} - \mathbf{U}_{(\tau)}\mathbf{x}_{(\tau-1)}\right\|_F^2 .$$

This decoding rule does not require CSI at the receiver.

By using the arguments in Chap. 2, we can show that when the set of code-matrices $\mathcal{U}$ is fully diverse, differential DSTC can achieve the same diversity order as DSTC, which is $R\left(1 - \log_e \log_e P / \log_e P\right)$. It can also be easily proved that compared with (coherent) DSTC, differential DSTC has a 3dB loss in performance due to the doubling of the noise power.

4.2 Code Design

The code design problem for differential DSTC is the design of the relay matrices $\mathbf{A}_i$'s, $\mathbf{B}_i$'s and the unitary matrix set $\mathcal{U}$ such that the condition in (4.3) is satisfied. In this section, several code designs are introduced, namely Alamouti code [1, 5], square real orthogonal code [5, 10], $Sp(2)$ code [4, 5], and circulant code [5].

4.2.1 Alamouti Code

For a relay network with two relay antennas, i.e., $R = 2$, Alamouti code [1] can be used to realize the differential DSTC transmission. Design the relay transformation matrices as

$$\mathbf{A}_1 = \mathbf{I}_2, \ \mathbf{A}_2 = \mathbf{0}, \ \mathbf{B}_1 = \mathbf{0}, \ \mathbf{B}_2 = \begin{bmatrix} 0 & -1 \\ 1 & 0 \end{bmatrix}. \tag{4.5}$$

It can be shown via direct matrix multiplication that a matrix $\mathbf{U}$ satisfies (4.3) if and only if it has the structure of Alamouti code:

$$\mathbf{U} = \begin{bmatrix} u_1 & -u_2^* \\ u_2 & u_1^* \end{bmatrix}.$$

Therefore, the set of unitary code-matrices is designed as

$$\mathcal{U} = \left\{ \frac{1}{\sqrt{|u_1|^2 + |u_2|^2}} \begin{bmatrix} u_1 & -u_2^* \\ u_2 & u_1^* \end{bmatrix} | u_i \in \mathcal{F}_i, \qquad \text{for } i = 1, 2. \right\},$$

where $\mathcal{F}_i$ is some finite set, for example, phase-shift keying (PSK) or quadrature-amplitude modulation (QAM) signals. If the cardinality of $\mathcal{F}_1$ and $\mathcal{F}_2$ are L_1 and L_2, respectively, the transmission rate of the Alamouti code is $\log_2(L_1 L_2)/(2T)$ bits per transmission.

4.2.2 Square Real Orthogonal Codes

Square real orthogonal codes were proposed in [10]. They only exist for dimensions two, four, and eight and can be used in networks with two, four, and eight relays. To help the presentation, define

$$\mathbf{T}_1 \triangleq \begin{bmatrix} 0 & -1 \\ 1 & 0 \end{bmatrix}, \mathbf{T}_2 \triangleq \begin{bmatrix} 1 & 0 \\ 0 & -1 \end{bmatrix}, \mathbf{T}_3 \triangleq \mathbf{T}_1 \mathbf{T}_2 = \begin{bmatrix} 0 & 1 \\ 1 & 0 \end{bmatrix}.$$

To use square real orthogonal codes for differential DSTC, we design the relay transformation matrices as follows. For networks with two relay antennas, let $\mathbf{B}_1 = \mathbf{B}_2 = \mathbf{0}$ and

$$\mathbf{A}_1 = \mathbf{I}_2, \qquad \mathbf{A}_2 = \mathbf{T}_1; \tag{4.6}$$

for networks with four relay antennas, let $\mathbf{B}_1 = \mathbf{B}_2 = \mathbf{B}_3 = \mathbf{B}_4 = \mathbf{0}$ and

$$\mathbf{A}_1 = \mathbf{I}_4, \ \mathbf{A}_2 = \begin{bmatrix} \mathbf{T}_1 & 0 \\ 0 & \mathbf{T}_1 \end{bmatrix}, \ \mathbf{A}_3 = \begin{bmatrix} 0 & -\mathbf{T}_2 \\ \mathbf{T}_2 & 0 \end{bmatrix}, \ \mathbf{A}_4 = \begin{bmatrix} 0 & -\mathbf{T}_3 \\ \mathbf{T}_3 & 0 \end{bmatrix}; \tag{4.7}$$

and for networks with eight relay antennas, let $\mathbf{B}_1 = \cdots = \mathbf{B}_8 = \mathbf{0}$ and

$$\mathbf{A}_1 = \mathbf{I}_8, \qquad\qquad \mathbf{A}_2 = \begin{bmatrix} \mathbf{T}_1 & 0 & 0 & 0 \\ 0 & -\mathbf{T}_1 & 0 & 0 \\ 0 & 0 & -\mathbf{T}_1 & 0 \\ 0 & 0 & 0 & \mathbf{T}_1 \end{bmatrix},$$

$$\mathbf{A}_3 = \begin{bmatrix} 0 & -\mathbf{I}_2 & 0 & 0 \\ \mathbf{I}_2 & 0 & 0 & 0 \\ 0 & 0 & 0 & \mathbf{I}_2 \\ 0 & 0 & -\mathbf{I}_2 & 0 \end{bmatrix}, \quad \mathbf{A}_4 = \begin{bmatrix} 0 & \mathbf{T}_2 & 0 & 0 \\ \mathbf{T}_2 & 0 & 0 & 0 \\ 0 & 0 & 0 & -\mathbf{T}_2 \\ 0 & 0 & -\mathbf{T}_2 & 0 \end{bmatrix},$$

$$\mathbf{A}_5 = \begin{bmatrix} 0 & 0 & -\mathbf{I}_2 & 0 \\ 0 & 0 & 0 & -\mathbf{I}_2 \\ \mathbf{I}_2 & 0 & 0 & 0 \\ 0 & \mathbf{I}_2 & 0 & 0 \end{bmatrix}, \quad \mathbf{A}_6 = \begin{bmatrix} 0 & 0 & \mathbf{T}_1 & 0 \\ 0 & 0 & 0 & \mathbf{T}_1 \\ \mathbf{T}_1 & 0 & 0 & 0 \\ 0 & \mathbf{T}_1 & 0 & 0 \end{bmatrix},$$

$$A_7 = \begin{bmatrix} 0 & 0 & 0 & -T_2 \\ 0 & 0 & T_2 & 0 \\ 0 & -T_2 & 0 & 0 \\ T_2 & 0 & 0 & 0 \end{bmatrix}, \quad A_8 = \begin{bmatrix} 0 & 0 & 0 & -T_3 \\ 0 & 0 & T_3 & 0 \\ 0 & -T_3 & 0 & 0 \\ T_3 & 0 & 0 & 0 \end{bmatrix}. \tag{4.8}$$

It can be proved by direct matrix multiplication that a real matrix commutes with the set $\{A_1, \ldots, A_R\}$ in (4.6), (4.7), and (4.8) if and only if it has the following real square orthogonal structure, respectively:

$$\begin{bmatrix} u_1 & -u_2 \\ u_2 & u_1 \end{bmatrix}, \tag{4.9}$$

$$\begin{bmatrix} u_1 & -u_2 & -u_3 & -u_4 \\ u_2 & u_1 & u_4 & -u_3 \\ u_3 & -u_4 & u_1 & u_2 \\ u_4 & u_3 & -u_2 & u_1 \end{bmatrix}, \tag{4.10}$$

$$\begin{bmatrix} u_1 & -u_2 & -u_3 & -u_4 & -u_5 & -u_6 & -u_7 & -u_8 \\ u_2 & u_1 & -u_4 & u_3 & -u_6 & u_5 & u_8 & -u_7 \\ u_3 & u_4 & u_1 & -u_2 & -u_7 & -u_8 & u_5 & u_6 \\ u_4 & -u_3 & u_2 & u_1 & -u_8 & u_7 & -u_6 & u_5 \\ u_5 & u_6 & u_7 & u_8 & u_1 & -u_2 & -u_3 & -u_4 \\ u_6 & -u_5 & u_8 & -u_7 & u_2 & u_1 & u_4 & -u_3 \\ u_7 & -u_8 & -u_5 & u_6 & u_3 & -u_4 & u_1 & u_2 \\ u_8 & u_7 & -u_6 & -u_5 & u_4 & u_3 & -u_2 & u_1 \end{bmatrix}. \tag{4.11}$$

Thus, the data-matrices should be designed to have the square real orthogonal structure in (4.9)–(4.11) with the information symbols u_i's selected from a real modulation such as pulse-amplitude modulation (PAM). The distributed space-time codewords generated at the receiver have the same square real orthogonal structure as the data-matrices.

4.2.3 Sp(2) Code

$Sp(2)$ code was proposed in [4] for differential space-time coding in multiple-antenna systems with four transmit antennas. It can be seen as an extension of Alamouti code to dimension four. Its symbol rate is 1. Each code-matrix of the code has the following structure:

$$U_{Sp(2)}(a_1, a_2, b_1, b_2) = \frac{1}{\sqrt{2}} \begin{bmatrix} V_1 V_2 & V_1 \overline{V_2} \\ -\overline{V_1} V_2 & \overline{V_1} \overline{V_2} \end{bmatrix},$$

where

$$V_i = \frac{1}{\sqrt{|a_i|^2 + |b_i|^2}} \begin{bmatrix} a_i & b_i \\ -b_i^* & a_i^* \end{bmatrix}.$$

It is straightforward to verify that $\mathbf{U}_{Sp(2)}$ is a unitary matrix for any a_1, a_2, a_3, a_4. Let $\mathcal{F}_1, \mathcal{F}_2, \mathcal{G}_1, \mathcal{G}_2$ be the constellation sets for the four symbols a_1, a_2, b_1, b_2, respectively. A $Sp(2)$ code can thus be represented as

$$\mathcal{U} = \left\{ \mathbf{U}_{Sp(2)}(a_1, a_2, b_1, b_2) \,|\, a_1 \in \mathcal{F}_1, a_2 \in \mathcal{F}_2, b_1 \in \mathcal{G}_1, b_2 \in \mathcal{G}_2 \right\}. \qquad (4.12)$$

Choices of $\mathcal{F}_i$ and $\mathcal{G}_i$ are arbitrary and are not constraint to be real.

To further understand the structure of $Sp(2)$ code, define

$$u_1 \triangleq \frac{a_1 a_2 - b_1 b_2^*}{\sqrt{2} \prod_{i=1}^{2} \sqrt{|a_i|^2 + |b_i|^2}}, \qquad u_2 \triangleq -\frac{a_1^* b_2^* + b_1^* a_2}{\sqrt{2} \prod_{i=1}^{2} \sqrt{|a_i|^2 + |b_i|^2}},$$
$$u_3 \triangleq -\frac{a_1^* a_2 - b_1^* b_2^*}{\sqrt{2} \prod_{i=1}^{2} \sqrt{|a_i|^2 + |b_i|^2}}, \qquad u_4 \triangleq \frac{a_1 b_2^* + b_1 a_2}{\sqrt{2} \prod_{i=1}^{2} \sqrt{|a_i|^2 + |b_i|^2}}. \qquad (4.13)$$

It can be shown using straightforward calculation that

$$\mathbf{U}_{Sp(2)}(a_1, a_2, b_1, b_2) = \begin{bmatrix} u_1 & -u_2^* & -u_3^* & u_4 \\ u_2 & u_1^* & -u_4^* & -u_3 \\ u_3 & -u_4^* & u_1^* & -u_2 \\ u_4 & u_3^* & u_2^* & u_1 \end{bmatrix}. \qquad (4.14)$$

This shows that $Sp(2)$ code-matrix is quasi-orthogonal [3, 11]. However, a $Sp(2)$ code-matrix differs to a quasi-orthogonal design in that a $Sp(2)$ code-matrix is unitary due to the special structure of u_1, u_2, u_3, u_4 in (4.13). A quasi-orthogonal matrix in general is not unitary. The special structure of u_1, u_2, u_3, u_4 are obtained from the analysis on the special unitary Lie group $Sp(2)$ [2, 9]. For differential transmission, e.g., differential space-time coding and differential DSTC, having the data-matrix unitary maintains the transmit power.

$Sp(2)$ code can be applied for differential DSTC in networks with four relay antennas. Design the data-matrix set as in (4.13), and design the transformation matrices used at the relay as

$$\mathbf{A}_1 = \mathbf{I}_4, \ \mathbf{A}_2 = \mathbf{0}, \ \mathbf{A}_3 = \mathbf{0}, \ \mathbf{A}_4 = \begin{bmatrix} \mathbf{0} & -\mathbf{T}_1 \\ \mathbf{T}_1 & \mathbf{0} \end{bmatrix},$$
$$\mathbf{B}_1 = \mathbf{0}, \ \mathbf{B}_2 = \begin{bmatrix} \mathbf{T}_1 & \mathbf{0} \\ \mathbf{0} & \mathbf{T}_1 \end{bmatrix}, \ \mathbf{B}_3 = \begin{bmatrix} \mathbf{0} & -\mathbf{I}_2 \\ \mathbf{I}_2 & \mathbf{0} \end{bmatrix}, \ \mathbf{B}_4 = \mathbf{0}. \qquad (4.15)$$

It can be shown via straightforward calculation that condition (4.3) is satisfied. The distributed space-time codewords formed at the receiver also have the quasi-orthogonal structure in (4.14). When $\mathcal{F}_1, \mathcal{F}_2, \mathcal{G}_1$, and $\mathcal{G}_2$ are chosen as PSK constellations, sufficient and necessary conditions for the $Sp(2)$ code to be fully diverse were provided in [4].

4.2.4 Circulant Code

Although Alamouti code, square real orthogonal code, and $Sp(2)$ code have good performance and low decoding complexity, they only work for networks with two, four, and eight relays. A type of circulant codes that work for networks with any number of relays was proposed in [5].

For a circulant code, the relay transformation matrices are designed as

$$\mathbf{A}_i = \mathbf{A}^{i-1}, \quad \mathbf{B}_i = \mathbf{0}, \tag{4.16}$$

where $\mathbf{A}$ is defined as

$$\mathbf{A} \triangleq \begin{bmatrix} 0 & 1 & 0 & \cdots & 0 \\ 0 & 0 & 1 & \cdots & 0 \\ \vdots & \vdots & \vdots & \ddots & \vdots \\ 1 & 0 & 0 & \cdots & 0 \end{bmatrix}. \tag{4.17}$$

The motivation of this design is from the following results in matrix theory: a matrix $\mathbf{M}_1$ commutes with all matrices that commute with $\mathbf{M}_2$ if and only if $\mathbf{M}_1$ is a polynomial of $\mathbf{M}_2$ [7]. With the $\mathbf{A}_i$ design in (4.16), we know that any matrix that commutes with $\mathbf{A}$ commutes with the set $\{\mathbf{A}_1, \ldots, \mathbf{A}_R\}$.

Condition (4.3) thus reduces to: for any $\mathbf{U} \in \mathcal{U}$, $\mathbf{UA} = \mathbf{AU}$. We need to find the set of $\mathbf{U}$ that commutes with $\mathbf{A}$. It is straightforward to prove that a matrix commutes with $\mathbf{A}$ defined in (4.17) if and only if it is a circulant matrix:

$$\mathbf{U} = \begin{bmatrix} u_1 & u_2 & u_3 & \cdots & u_R \\ u_R & u_1 & u_2 & \cdots & u_{R-1} \\ u_{R-1} & u_R & u_1 & \cdots & u_{R-2} \\ \vdots & \vdots & \vdots & \ddots & \vdots \\ u_2 & u_3 & u_4 & \cdots & u_1 \end{bmatrix}.$$

However, in general, a circulant matrix is not unitary. To obtain unitarity, the following set of special circulant matrices are used as code-matrices:

$$\mathcal{U} = \left\{ u_1 \mathbf{I}, u_2 \mathbf{A}, \ldots, u_R \mathbf{A}^{R-1} | u_i \in \mathcal{F}_i, \text{ for } i = 1, 2, \ldots, R. \right\}, \tag{4.18}$$

where $\mathbf{A}$ is defined in (4.17) and $\mathcal{F}_i$'s are constellations. For the code-matrix to be unitary, elements in $\mathcal{F}_i$ must be have unit-norm. The bit rate of this code can be calculated to be $\frac{1}{2T} \log_2 \sum_{i=1}^{R} |\mathcal{F}_i|$, where $|\mathcal{F}_i|$ is the cardinality of $\mathcal{F}_i$.

In what follows, two examples of circulant code are provided.

Example 4.1. [5] For a network with three relays, i.e., $R = 3$, design the transformation matrices at the relays as

$$\left\{ \mathbf{I}_3, \begin{bmatrix} 0 & 1 & 0 \\ 0 & 0 & 1 \\ 1 & 0 & 0 \end{bmatrix}, \begin{bmatrix} 0 & 0 & 1 \\ 1 & 0 & 0 \\ 0 & 1 & 0 \end{bmatrix} \right\}$$

and the set of data-matrices as

$$\mathcal{U} = \left\{ \begin{bmatrix} u_1 & 0 & 0 \\ 0 & u_1 & 0 \\ 0 & 0 & u_1 \end{bmatrix}, \begin{bmatrix} 0 & u_2 & 0 \\ 0 & 0 & u_2 \\ u_2 & 0 & 0 \end{bmatrix}, \begin{bmatrix} 0 & 0 & u_3 \\ u_3 & 0 & 0 \\ 0 & u_3 & 0 \end{bmatrix} \Big| u_i \in \mathcal{F}_i, i = 1, 2, 3. \right\}.$$

If $\mathcal{F}_i$ is chosen as quadrature phase-shift keying (QPSK), the cardinality of $\mathcal{U}$ is 12. The bit rate is therefore $(\log_2 12)/6 = 0.5975$ bit per transmission.

Example 4.2. For a network with five relays, i.e., $T = R = 5$, let

$$\mathbf{A} = \begin{bmatrix} 0 & 1 & 0 & 0 & 0 \\ 0 & 0 & 1 & 0 & 0 \\ 0 & 0 & 0 & 1 & 0 \\ 0 & 0 & 0 & 0 & 1 \\ 1 & 0 & 0 & 0 & 0 \end{bmatrix}.$$

Design matrices used at relays as $\{\mathbf{I}_5, \mathbf{A}, \mathbf{A}^2, \mathbf{A}^3, \mathbf{A}^4\}$ and the set of data matrices as in (4.18). $\mathcal{F}_1, \mathcal{F}_2, \mathcal{F}_3$ are chosen as QPSK and $\mathcal{F}_4, \mathcal{F}_5$ are chosen as 8-PSK. The cardinality of $\mathcal{U}$ is thus 28. The bit rate of this code is $(\log_2 28)/10 = 0.4807$ bit per transmission.

For a circulant code to have full diversity, the constellation sets need to be designed carefully. When $\mathcal{F}_i$'s are chosen as M-PSK rotated by an angle θ_i, i.e.,

$$\mathcal{F}_i = \left\{ e^{j\theta_i}, \ldots, e^{j\left(2\pi \frac{M-1}{M} + \theta_i\right)} \right\}. \tag{4.19}$$

the necessary and sufficient condition for the full diversity of circulant code has been provided in [5]. The result is stated in the following theorem.

Theorem 4.1. *[5] Design $\mathcal{F}_i$ as in (4.19). The circulant code in (4.18) is fully diverse if any only if*

$$\frac{\theta_{i_1} - \theta_{i_2}}{2\pi} \mathrm{lcm}\left(\frac{R}{\gcd(R, i_2 - i_1)}, M \right)$$

is not an integer for all $1 \le i_1 < i_2 \le R$, where $\mathrm{lcm}(m, n)$ and $\gcd(m, n)$ are the least common multiplier and the greatest common divider of integers m and n, respectively.

For $R = 2$ and small M, the optimal angles that result in the largest coding gain were found analytically in [5]. For larger R and M, the optimal angles can always be found by off-line numerical search.

4.3 Simulation on Error Probability

In this section, simulated block error rates of differential DSTC for several network scenarios are demonstrated.

The first to consider is a network with two relay antennas, i.e., $R = 2$. We set $T = 2$ and use an Alamouti code whose information symbols are modulated as BPSK and 8PSK. Figure 4.1 shows that the diversity order of the network is close to 2. The code with BPSK (whose transmission rate is 1/2 bit per transmission) is about 10dB better than that with 8PSK (whose transmission rate is 3/2 bit per transmission). Compared with the corresponding coherent DSTC with the same codes, differential DSTC is about 3dB worse. But it requires no CSI at any node of the network.

Next, networks with 3, 4, and 5 relays using circulant codes with QPSK information symbols are considered. Thus, the transmission rates of the three networks are 0.60, 0.50, and 0.43 bit per transmission, respectively. The optimal angles of the QPSK rotations given in [5] are used. Figure 4.2 shows that the diversity order of each network is approximately R.

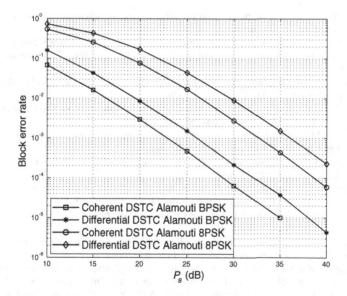

Fig. 4.1 Performance of differential DSTC for a network with $T = R = 2$ and Alamouti code. BPSK and 8PSK are used

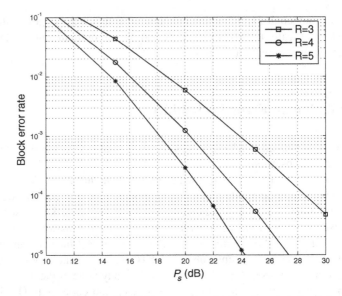

Fig. 4.2 Performance of differential DSTC for networks with $T = R = 3, 4, 5$ and circulant codes under QPSK modulation

References

1. Alamouti SM (1998) A simple transmit diversity scheme for wireless communications. IEEE J on Selected Areas in Communications, 16:1451–1458.
2. Bröcker T and tom Dieck T (1995) Representations of compact Lie Groups. Springer.
3. Jafarkhani H (2001) A quasi-orthogonal space-time block codes. IEEE T on Communications, 49:1–4.
4. Jing Y and Hassibi B (2004) Design of fully-diverse multiple-antenna codes based on $Sp(2)$. IEEE T on Information Theory, 50:2639–2656.
5. Jing Y and Jafarkhani H (2008) Distributed differential space-time coding in wireless relay networks. IEEE T on Communications, 56:1092–1100.
6. Kiran T and Rajan SB (2006) Partial-coherent distributed space-time codes with differential encoder and decoder. IEEE International S Information Theory, 547551.
7. Lagerstrom P (1945) A proof of a theorem on commutative matrices. Bulletin of the American Mathematical Society, 51:535–536.
8. Oggier F and Hassibi B (2006) A coding strategy for wireless networks with no channel information. Allerton C Communications, Control, and, Computing, 113–117.
9. Sattinger DH and Weaver OL (1986) Lie Groups and Algebras with applications to physics, geometry, and mechanics. Springer.
10. Tarokh V, Jafarkhani H, and Calderbank AR (1999) Space-time block codes from orthogonal designs. IEEE T on Information Theory, 45:1456–1467.
11. Wang H and Xia X-G (2005) On optimal quasi-orthogonal space-time block codes with minimum decoding complexity. IEEE International Symposium on Information Theory, 1168–1172.
12. Weisstein EW, Circulant determinant. MathWorldA Wolfram Web Resource, http://mathworld.wolfram.com/CirculantDeterminant. html

Chapter 5
Training and Training-Based Distributed Space-Time Coding

Abstract The coherent distributed space-time coding (DSTC) schemes introduced in Chaps. 2 and 3 require full channel state information (CSI) at the receiver. In reality, training and channel estimations need to be conducted to obtain the required CSI at the receiver. The estimated CSI is then used in the data transmission. This chapter is on channel training and training-based DSTC. First, the training and estimation of the global and individual channels of the relay network are considered in Sect. 5.1. Then, training-based DSTC, i.e, DSTC with estimated CSI, is studied in Sect. 5.2. After that, the training and estimation of the end-to-end channels for single-antenna relay network and multiple-antenna relay network are explained in Sects. 5.3 and 5.4, respectively.

5.1 Estimation of the Individual Channels at the Receiver

This section studies the estimation of both the channels from the relays to the receiver and the channels from the transmitter to the relays individually, at the receiver. We also call it the global channel state information (CSI) estimation at the receiver. If the receiver has global CSI, it can conduct the decoding of coherent DSTC. The general multiple-antenna multiple-relay network is studied directly. Results for single-antenna multiple-relay network can be obtained straightforwardly by setting the numbers of transmit and receiver antennas to 1. Materials in this section mainly follow the work in [10, 11].

Consider the general multiple-input-multiple-output (MIMO) relay network shown in Fig. 3.1, where there are M transmit antennas, a total of R relay antennas, and N receive antennas. The notation follow Sect. 3.1. $\mathbf{f}$ is the $MR \times 1$ channel vector from the transmitter to the relay antennas, also called *the TX-Relay channel vector* in short. $\mathbf{G}$ is the $R \times N$ channel matrix from the relay antennas to the receiver, also called *the Relay-RX channel matrix* in short. Their detailed definitions can be found in (3.1).

Y. Jing, *Distributed Space-Time Coding*, SpringerBriefs in Computer Science,
DOI: 10.1007/978-1-4614-6831-8_5, © The Author(s) 2013

This section is organized as follows. First, the training protocol is explained in Sect. 5.1.1. Then the estimation of the Relay-RX channels is demonstrated in Sect. 5.1.2. After that, the estimation of the TX-Relay channels under perfect and estimated Relay-RX channel information are given in Sects. 5.1.3 and 5.1.4, respectively.

Before getting into the details of the training and estimation. We explain the notation used in this chapter. For the entire chapter, P_s is used for the average transmit power of the transmitter for both training and data transmission. P_r is used for the average transmit power of the relay for both training and data transmission.[1] The subscript $(\cdot)_p$ is used to indicate that a symbol is for the training interval. This applies to symbols representing the signals (e.g., pilot signals, transmitted signals, and received signals at any node of the network), noises (relay noises, receiver noises, equivalent noises), and time durations (training time duration for different training stages), Symbols without $(\cdot)_p$ are for the data-transmission interval. Symbols with subscript $(\cdot)_r$ are for the relays and symbols with subscript $(\cdot)_d$ are for the receiver.

5.1.1 Training Design

The training phase has two stages, where the first stage is for the training of $\mathbf{G}$, the Relay-RX channel matrix; and the second stage is for the training of $\mathbf{f}$, the TX-Relay channel vector.

Stage 1: Training of the Relay-RX Channels

The Relay-RX link can be seen as a virtual $R \times N$ point-to-point multiple-antenna system, whose training has been well investigated [2]. Thus, for the training of the Relay-TX channels, we follow the work in [2].

Denote the length of this training stage as $T_{p,\mathbf{G}}$. A $T_{p,\mathbf{G}} \times R$ pilot matrix $\mathbf{U}_p$ is transmitted by the relays. That is, during the $T_{p,\mathbf{G}}$ transmission slots, the ith relay antenna sends the ith column of $\mathbf{U}_p$. The signal matrix the receiver receives, denoted as $\mathbf{Y}_p$, satisfies

$$\mathbf{Y}_p = \sqrt{P_r T_{p,\mathbf{G}}} \mathbf{U}_p \mathbf{G} + \mathbf{N}_{d,\mathbf{G},p}, \qquad (5.1)$$

where P_r is the average transmit power of each relay antenna, and $\mathbf{N}_{d,p,\mathbf{G}}$ is the $T_{p,\mathbf{G}} \times N$ noise matrix at the receiver. The subscripts d, $\mathbf{G}$, p are used to indicate that the noise is at the receiver, for the training of $\mathbf{G}$, and for the training stage. From the results in [2], for good estimation quality, $T_{p,\mathbf{G}} \geq R$ is required to have

---

[1] This implies that for both the training and the data-transmission intervals, the transmitter and relays are assumed to use the same power. The optimal power allocations (including the power allocations between the transmitter and relays during training, between different training steps, and between the training interval and data transmission interval) are not considered. This chapter only concerns with the training and channel estimation schemes.

the number of independent training equations in (5.1) no smaller than the number of channel coefficients to be estimated. To minimize the power of the estimation error, $\mathbf{U}_p$ should be unitary, i.e., $\mathbf{U}_p^*\mathbf{U}_p = \mathbf{I}_R$.

Stage 2: Training of the TX-Relay Channels

Stage 2 of the training phase is for the training of $\mathbf{f}$, the TX-Relay channel vector. The two-step distributed space-time coding (DSTC) scheme is used. Denote the length of this training stage as $2T_p$, where each step of the DSTC takes T_p time slots.

Let $\mathbf{B}_p$ be the $T_p \times M$ pilot matrix sent by the transmitter, normalized as $\mathrm{tr}(\mathbf{B}_p^*\mathbf{B}_p) = M$. Let $\mathbf{A}_{1,p}, \cdots, \mathbf{A}_{R,p}$ be the $T_p \times T_p$ unitary transformation matrices used at the relay antennas during training. Let $\mathbf{n}_{r,i,p}$ be the noise vector at the ith relay antenna and $\mathbf{N}_{d,p}$ be the noise matrix at the receiver. Recall that the powers used at the transmitter and each relay antenna are denoted as P_s and P_r, respectively. Define

$$\alpha_p \triangleq \frac{P_r}{1 + P_s}, \quad \beta_p \triangleq \sqrt{\frac{P_s P_r T_p}{M(P_s + 1)}}, \tag{5.2}$$

$$\tilde{\mathbf{S}}_p \triangleq \mathrm{diag}\{\mathbf{A}_{1,p}\mathbf{B}_p, \cdots, \mathbf{A}_{R,p}\mathbf{B}_p\}, \tag{5.3}$$

$$\mathbf{Z}_p \triangleq (\mathbf{G}^t \otimes \mathbf{I}_{T_p})\tilde{\mathbf{S}}_p. \tag{5.4}$$

Note that α_p in (5.3) is the same as α, firstly defined in (2.16). This is because we assume that the transmitter and relay powers are the same for training and data-transmission. Denote the received signal matrix at the receiver as $\mathbf{X}_p$. Following the DSTC description in Sect. 3.2, the training equation can be written as:

$$\mathrm{vec}(\mathbf{X}_p) = \sqrt{\beta_p}\left[(\mathbf{G}^t \otimes \mathbf{I}_{T_p})\tilde{\mathbf{S}}_p\right]\mathbf{f} + \mathrm{vec}(\mathbf{W}_p) = \sqrt{\beta_p}\mathbf{Z}_p\mathbf{f} + \mathrm{vec}(\mathbf{W}_p), \tag{5.5}$$

where $\mathbf{W}_p$ is the equivalent noise, given as

$$\mathbf{W}_p \triangleq \sqrt{\alpha_p}\left[\mathbf{A}_{1,p}\mathbf{n}_{r,1,p} \quad \cdots \quad \mathbf{A}_{r,R,p}\mathbf{n}_{r,R,p}\right]\mathbf{G} + \mathbf{N}_{d,p}.$$

Also, following the results in Sect. 3.1, the covariance matrix of the equivalent noise vector $\mathrm{vec}(\mathbf{W}_p)$ is

$$\mathbf{R}_{\mathrm{vec}(\mathbf{W}_p)} \triangleq \left(\mathbf{I}_N + \alpha_p\overline{\mathbf{G}^*\mathbf{G}}\right) \otimes \mathbf{I}_{T_p}. \tag{5.6}$$

5.1.2 Estimation of the Relay-RX Channels

We can use the observation $\mathbf{Y}_p$, obtained in the first stage of training, to estimation the Relay-RX channels. From (5.1), we see that the observation model is a Gaussian one. The minimum-mean-square-error (MMSE) estimate is [6]

$$\hat{\mathbf{G}} = \sqrt{\frac{1}{P_r T_{p,G}}} \left(\frac{1}{P_r T_{p,G}} \mathbf{I}_R + \mathbf{U}_p^* \mathbf{U}_p \right)^{-1} \mathbf{U}_p^* \mathbf{Y}_p. \tag{5.7}$$

Let $\Delta \mathbf{g} \triangleq \text{vec}(\mathbf{G} - \hat{\mathbf{G}})$, which is the vector of estimation error on $\mathbf{G}$. With a Gaussian observation model, the error vector $\Delta \mathbf{g}$ has been shown to be a Gaussian random vector. Its mean can be straightforwardly shown to be zero. Its covariance matrix is

$$\mathbf{R}_{\Delta \mathbf{g}} \triangleq \mathbb{E}\{(\Delta \mathbf{g})^*(\Delta \mathbf{g})\} = \frac{1}{1 + P_r T_{p,G}} \mathbf{I}_{NR}.$$

The power of the estimation error on $\mathbf{G}$ is thus

$$\text{tr}(\mathbf{R}_{\Delta \mathbf{g}}) = \frac{R}{1 + P_r T_{p,G}} = \mathcal{O}\left(\frac{1}{P_r} \right).$$

5.1.3 Estimation of the TX-Relay Channels with Perfect Relay-RX Channel Information

The main challenge in the estimation of the global CSI at the receiver actually lies in the estimation of $\mathbf{f}$, the TX-Relay channel vector. Notice that the receiver is actually not directly connected with the channels in $\mathbf{f}$.

In this section, the ideal case that $\mathbf{G}$ is known perfectly at the receiver is considered. In reality, only the estimation on $\mathbf{G}$ is known. However, as shown in the previous subsection, the estimation error on $\mathbf{G}$ scales as $1/P_r$. Thus, when the relay transmit power is high enough, the estimation error on $\mathbf{G}$ is negligible. The estimation of $\mathbf{f}$ under imperfect $\mathbf{G}$ will be considered later in Sect. 5.1.4.

Estimation Rule

From (5.5), we can see that when $\mathbf{G}$ is known, $\mathbf{Z}_p$ can function as the training matrix in estimating $\mathbf{f}$. If $\mathbf{G}$ is perfectly known, the observation model is a traditional Gaussian model. The linear minimum-mean-square-error (LMMSE) estimation of $\mathbf{f}$, which is also the MMSE estimation, can be straightforwardly obtained using Bayesian Gauss-Markov theorem [6, 10] to be

$$\hat{\mathbf{f}} = \sqrt{\beta_p} \left(\mathbf{I}_{MR} + \beta_p \mathbf{Z}_p^* \mathbf{R}_{\text{vec}(\mathbf{W}_p)}^{-1} \mathbf{Z}_p \right)^{-1} \mathbf{Z}_p^* \mathbf{R}_{\text{vec}(\mathbf{W}_p)}^{-1} \text{vec}(\mathbf{X}_p). \tag{5.8}$$

Note that $\mathbf{R}_{\text{vec}(\mathbf{W}_p)}$ is given in (5.6).

Let $\Delta \mathbf{f} \triangleq \mathbf{f} - \hat{\mathbf{f}}$, which is the error vector of the estimation on $\mathbf{f}$. It is Gaussian distributed. Its mean can be shown straightforwardly to be zero. Its covariance matrix is

$$\mathbf{R}_{\Delta f} \triangleq \text{Cov}(\Delta \mathbf{f}) = \left(\mathbf{I}_{MR} + \beta_p^2 \mathbf{Z}_p^* \mathbf{R}_{\text{vec}(\mathbf{W}_p)}^{-1} \mathbf{Z}_p \right)^{-1}.$$

Pilot Design

The pilot design problem is the design of both $\mathbf{B}_p$ (the matrix sent by the transmitter) and $\mathbf{A}_{i,p}$'s (the relay transformation matrices) that minimize the power of the estimation error. It can be represented as

$$\min_{\mathbf{A}_{i,p}, \mathbf{B}_p} \ \text{tr}\,(\mathbf{R}_{\Delta f})$$

$$\text{s.t.}\ \text{tr}(\mathbf{B}_p^* \mathbf{B}_p) = M,$$

$$\text{and } \mathbf{A}_{i,p}^* \mathbf{A}_{i,p} = \mathbf{I}_{T_p} \quad \text{for } i = 1, \dots, R. \tag{5.9}$$

The pilot design was solved in [10], where the cases of $T_p \geq MR$ and $T_p < MR$ were considered separately. The result for the first case is represented in the following theorem.

Theorem 5.1. *[10] Let* $\mathbf{S}_p \triangleq [\mathbf{A}_{1,p}\mathbf{B}_p \ \cdots \ \mathbf{A}_{R,p}\mathbf{B}_p]$. *When* $T_p \geq MR$, *the power of the estimation error on* $\mathbf{f}$ *is minimized when* $\mathbf{S}_p^* \mathbf{S}_p = \mathbf{I}_{MR}$.

Note that $\mathbf{S}_p$ is the pilot distributed space-time codeword of the second training stage. Theorem 1 says that when $T_p \geq MR$, the optimal pilot design is to make the distributed space-time codeword $\mathbf{S}_p$ unitary, which is consistent with the pilot design in point-to-point MIMO systems. The code design problem thus becomes finding $\mathbf{B}_p$ and $\mathbf{A}_{i,p}$'s such that $\mathbf{S}_p$ is a $T_p \times MR$ unitary matrix. An algorithm for such optimal pilot design is given below [10].

Algorithm 1 [10] Optimal pilot design for $T_p \geq MR$.

1: Let $\mathbf{B}_p = \begin{bmatrix} \mathbf{I}_M & \mathbf{0}_{M,T_p-M} \end{bmatrix}^t$.
2: Generate a $T_p \times MR$ unitary matrix $\mathbf{S}_{po}$ (e.g., $\mathbf{S}_{po} = \begin{bmatrix} \mathbf{I}_{MR} & \mathbf{0}_{MR,T_p-MR} \end{bmatrix}^t$).
3: Generate the $MR \times MR$ permutation matrix $\mathbf{U}_i$ by switching the first M columns with the $[(i-1)M + 1]$th, $\cdots$, (iM)th columns of the $MR \times MR$ identity matrix. Let $\mathbf{A}_{i,p} = \begin{bmatrix} \mathbf{S}_{po}\mathbf{U}_i & (\mathbf{S}_{po}\mathbf{U}_i)^{\perp} \end{bmatrix}$.

Note that $T_p \geq MR \geq M$, so Step 1 of Algorithm 1 is always valid. Divide $\mathbf{S}_{po}$ into R blocks each with dimension $T_p \times M$: $\mathbf{S}_{po} = [\mathbf{S}_{p1} \ \cdots \ \mathbf{S}_{pR}]$. From Step 3, the effect of right-multiplying $\mathbf{S}_{po}$ with $\mathbf{U}_i$ is to switch $\mathbf{S}_{p1}$ and $\mathbf{S}_{pi}$, making $\mathbf{A}_{i,p} = \begin{bmatrix} \mathbf{S}_{pi} & \mathbf{S}_{pi}^{\perp} \end{bmatrix}$, where $\mathbf{S}_{pi}^{\perp}$ is the orthogonal complement of $\mathbf{S}_{pi}$. With the $\mathbf{B}_p$ generated in Step 1, it follows that

$$\mathbf{A}_{i,p}\mathbf{B}_p = \begin{bmatrix} \mathbf{S}_{pi} & \mathbf{S}_{pi}^\perp \end{bmatrix} \begin{bmatrix} \mathbf{I}_M & \mathbf{0}_{M,N_p-M} \end{bmatrix}^t = \mathbf{S}_{pi}.$$

Hence,

$$\mathbf{S}_p = \begin{bmatrix} \mathbf{A}_{1,p}\mathbf{B}_p & \cdots & \mathbf{A}_{R,p}\mathbf{B}_p \end{bmatrix} = \begin{bmatrix} \mathbf{S}_{p1} & \cdots & \mathbf{S}_{pR} \end{bmatrix} = \mathbf{S}_{po}$$

which is unitary. Thus, Algorithm 1 provides an optimal pilot design.

When $T_p < MR$, $\mathbf{S}_p^*\mathbf{S}_p$ is not full rank and $\mathbf{S}_p^*\mathbf{S}_p = \mathbf{I}_{MR}$ cannot be satisfied. A suboptimal pilot design was proposed in [10]. Consider a smaller network with $\tilde{R}$ relay antennas, where $\tilde{R} \triangleq \max\{1, \lfloor T_p/M \rfloor\}$. Note that $\tilde{R} \leq R$. For this smaller network, $T_p \geq M\tilde{R}$ is satisfied. Using Algorithm 1, an optimal pilot design for the smaller network is possible, denoted as $\mathbf{B}_p, \mathbf{A}_{1,p}, \cdots, \mathbf{A}_{\tilde{R},p}$. Let $\mathbf{S}_p' \triangleq \begin{bmatrix} \mathbf{A}_{1,p}\mathbf{B}_p & \cdots & \mathbf{A}_{\tilde{R},p}\mathbf{B}_p \end{bmatrix}$, which is the pilot space-time codeword for the smaller network. For the pilot design of the original larger network, the same $\mathbf{B}_p$ is used for the transmitter, while the transformation matrix at the ith relay antenna is designed to be $\mathbf{A}_{\mathrm{mod}\,(i-1,\tilde{R})+1,p}$. With this design, the pilot distributed space-time codeword for the original network has the following structure:

$$\mathbf{S}_p = \begin{bmatrix} \mathbf{S}_p' & \mathbf{S}_p' & \cdots \end{bmatrix}.$$

It was shown in [10] that this pilot design minimizes an upper bound on the power of the estimation error.

Estimation Error

In this part, the power of the estimation error is investigated, specifically, the scaling order of the power of the estimation error with respect to the training power. Assume that P_s, P_r (the transmit powers of the transmitter and each relay antenna) have the same order, i.e., $P_s, P_r \sim \mathcal{O}(P)$. The power of the estimation error, which is also the mean-square-error (MSE) of the estimation on $\mathbf{f}$, denoted as MSE($\mathbf{f}$), is the trace of the covariance matrix of $\Delta\mathbf{f}$, i.e., MSE($\mathbf{f}$) $= \mathbb{E}\{\mathrm{tr}(\mathbf{R}_{\Delta\mathbf{f}})\}$.

The calculation of MSE($\mathbf{f}$) is difficult for a general multiple-antenna multiple-relay network since the network is a complex concatenation of two virtual MIMO systems. In [9, 11], the following theorem on MSE($\mathbf{f}$) for the special case of $M = 1$, i.e., the transmitter has a single antenna, was derived.

Theorem 5.2. *[9, 11] For a relay network with single transmit antenna, R relay antennas, and N receive antennas, consider the LMMSE estimation of $\mathbf{f}$ in (5.8) and the aforementioned pilot design. When $T_p = 1$, MSE($\mathbf{f}$) $= R^2 \log_e P/P + \mathcal{O}(1/P)$; when $T_p \geq R$, MSE($\mathbf{f}$) $= \mathcal{O}(1/P)$.*

Theorem 5.2 says that if the training time is long enough ($T_p \geq R$), the MSE of the proposed estimation scales as $\mathcal{O}(1/P)$. If $T_p = 1$, there is an extra $\log_e P$ factor in the MSE. Intuitively and via simulation, having the power of the estimation error scales as $\mathcal{O}(1/P)$ is essential for full diversity in the data-transmission. When MSE($\mathbf{f}$) has a higher scaling than $\mathcal{O}(1/P)$, even with a small (compared with P) $\log_e P$ factor, a loss in diversity order may occur. More details will be discussed in the next section.

This theorem is for networks with single antenna at the transmitter only. For a general network whose transmitter can have multiple antennas, note that the channels connected to different transmit antennas are independent. We can conduct the estimation of the channels between each transmit antenna to the relay antennas separately. For example, one can first estimate the channels between the first transmit antenna and the relay antennas, then the channels between the second transmit antenna and the relay antennas, so on so forth, and finally the channels between the last transmit antenna and the relay antennas. Thus, with respect to the estimation of the TX-Relay channels, the general multiple-antenna multiple-relay network with M antennas at the transmitter can be seen as M independent networks with single antenna at the transmitter. With this observation, the results in Theorem 5.2 can be straightforwardly extended to obtain the following corollary for the general multiple-antenna multiple-relay network.

Corollary 5.1. *For a multiple-antenna relay network with M transmit antennas, R relay antennas, and N receive antennas, if $T_p \geq MR$, MSE($\mathbf{f}$) $= \mathcal{O}(1/P)$ can be achieved using the LMMSE estimation of $\mathbf{f}$ in (5.8) and the aforementioned pilot design.*

5.1.4 Estimation of the TX-Relay Channels with Estimated Relay-RX Channel Information

In reality, only an imperfect estimation of the Relay-RX channel matrix $\hat{\mathbf{G}}$ (obtained in the first training stage) is known. In this more realistic case, to estimate $\mathbf{f}$, a straightforward way is to use $\hat{\mathbf{G}}$ in (5.8). A more careful method is to take into consideration the estimation error $\Delta\mathbf{G} \triangleq \mathbf{G} - \hat{\mathbf{G}}$.

Estimation Rule

Replacing $\mathbf{G}$ with $\hat{\mathbf{G}} + \Delta\mathbf{G}$ in (5.5) leads to

$$\text{vec}(\mathbf{X}_p) = \sqrt{\beta_p}\hat{\mathbf{Z}}_p\mathbf{f} + \sqrt{\beta_p}\left[(\Delta\mathbf{G}^t \otimes \mathbf{I}_{T_p})\tilde{\mathbf{S}}_p\right]\mathbf{f} + \text{vec}(\mathbf{W}_{1,p} + \mathbf{W}_{2,p} + \mathbf{N}_{d,p}),$$

$$(5.10)$$

where

$$\hat{\mathbf{Z}}_p \triangleq \left(\hat{\mathbf{G}}^t \otimes \mathbf{I}_{T_p}\right)\tilde{\mathbf{S}}_p,$$

$$(5.11)$$

and

$$\mathbf{W}_{1,p} \triangleq \sqrt{\alpha_p}\left[\mathbf{A}_{1,p}\mathbf{n}_{r,1,p} \cdots \mathbf{A}_{R,p}\mathbf{n}_{r,R,p}\right]\hat{\mathbf{G}},$$

$$\mathbf{W}_{2,p} \triangleq \sqrt{\alpha_p} \left[\mathbf{A}_{1,p} \mathbf{n}_{r,1,p} \cdots \mathbf{A}_{R,p} \mathbf{n}_{r,R,p} \right] \Delta \mathbf{G}.$$

Recall that $\mathbf{n}_{r,i,p}$ is the noise vector at the ith relay antenna and $\mathbf{N}_{d,p}$ is the noise matrix at the receiver for the second training stage. Define

$$\mathbf{w}_{e,p} \triangleq \sqrt{\beta_p} \left[\left(\Delta \mathbf{G}^t \otimes \mathbf{I}_{T_p} \right) \tilde{\mathbf{S}}_p \right] \mathbf{f} + \text{vec}(\mathbf{W}_{1,p} + \mathbf{W}_{2,p} + \mathbf{N}_{d,p}),$$

which is the equivalent noise in the training equation. It has zero-mean. Its covariance matrix can be calculated as follows.

$$\begin{aligned}
\mathbf{R}_{\mathbf{w}_{e,p}} &\triangleq \mathbb{E}\{\mathbf{w}_{e,p} \mathbf{w}_{e,p}^*\} \\
&= \mathbb{E}\left\{ \beta_p \left[\left(\Delta \mathbf{G}^t \otimes \mathbf{I}_{T_p} \right) \tilde{\mathbf{S}}_p \right] \mathbf{f}\mathbf{f}^* \left[\left(\Delta \mathbf{G}^t \otimes \mathbf{I}_{T_p} \right) \tilde{\mathbf{S}}_p \right]^* + \text{vec}\left(\mathbf{W}_{1,p}\right) \text{vec}\left(\mathbf{W}_{1,p}\right)^* \right. \\
&\quad \left. + \text{vec}(\mathbf{W}_{2,p}) \text{vec}(\mathbf{W}_{2,p})^* + \text{vec}(\mathbf{N}_{d,p}) \text{vec}(\mathbf{N}_{d,p})^* \right\} \\
&= \frac{\beta_p}{1 + P_s T_{p,\mathbf{G}}} \left(\mathbf{I}_N \otimes \mathbf{S}_p \mathbf{S}_p^* \right) + \frac{R\alpha_p}{1 + P_s T_{p,\mathbf{G}}} \mathbf{I}_{NT_p} + \left(\mathbf{I}_N + \alpha_p \hat{\mathbf{G}}^t \overline{\hat{\mathbf{G}}} \right) \otimes \mathbf{I}_{T_p}. \quad (5.12)
\end{aligned}$$

The second equality is derived since $\left[\left(\Delta \mathbf{G}^t \otimes \mathbf{I}_{N_p} \right) \tilde{\mathbf{S}}_p \right] \mathbf{f}, \mathbf{W}_{1,p}, \mathbf{W}_{2,p}, \mathbf{N}_{d,p}$ are mutually uncorrelated.

Note that $\mathbf{w}_{e,p}$ depends on $\mathbf{f}$, which is the vector to be estimated. But this introduces no trouble in the LMMSE estimator since $\mathbf{w}_{e,p}$ is uncorrelated with $\mathbf{f}$. This can be seen by verifying that $\mathbb{E}(\mathbf{w}_{e,p}\mathbf{f}^*) = \mathbb{E}(\mathbf{w}_{e,p})\mathbb{E}(\mathbf{f}^*) = \mathbf{0}$. By Bayesian Gauss-Markov theorem [6], the LMMSE estimation of $\mathbf{f}$ is

$$\hat{\mathbf{f}} = \sqrt{\beta_p} \left(\mathbf{I}_{MR} + \beta_p \hat{\mathbf{Z}}_p^* \mathbf{R}_{\mathbf{w}_{e,p}}^{-1} \hat{\mathbf{Z}}_p \right)^{-1} \hat{\mathbf{Z}}_p^* \mathbf{R}_{\mathbf{w}_{e,p}}^{-1} \text{vec}(\mathbf{X}_p), \quad (5.13)$$

where $\mathbf{Z}_p$ is defined in (5.11) and $\mathbf{R}_{\mathbf{w}_{e,p}}$ is defined in (5.12).

Let $\Delta \mathbf{f} \triangleq \mathbf{f} - \hat{\mathbf{f}}$, which is the error vector of the estimation on $\mathbf{f}$. $\Delta \mathbf{f}$ has zero mean, i.e., $\mathbb{E}(\Delta \mathbf{f}) = \mathbf{0}$. Its covariance matrix can be calculated using Bayesian Gauss-Markov theorem to be

$$\mathbf{R}_{\Delta \mathbf{f}} \triangleq \left(\mathbf{I}_{MR} + \beta_p \hat{\mathbf{Z}}_p^* \mathbf{R}_{\mathbf{w}_{e,p}}^{-1} \hat{\mathbf{Z}}_p \right)^{-1}. \quad (5.14)$$

Notice that with the observation model in (5.10), $\mathbf{f}$ and $\text{vec}(\mathbf{X}_p)$ are not jointly Gaussian.[2] The LMMSE estimation of $\mathbf{f}$ in (5.13) is not the MMSE estimation, and the estimation error vector $\Delta \mathbf{f}$ is not Gaussian. But as the training power in the first training stage approaches infinity, i.e., $P_s \rightarrow \infty$, the estimation on $\mathbf{G}$ approaches perfect, in other words, $\hat{\mathbf{G}} \rightarrow \mathbf{G}$ with probability 1; and (5.13) reduces to (5.8), which is the MMSE estimation under perfect information on $\mathbf{G}$. Thus, the estimation in (5.13) approximates the MMSE estimation and $\Delta \mathbf{f}$ is approximately Gaussian with asymptotically high training power in the first training stage.

---

[2] This can be proved by realizing that $\text{vec}(\mathbf{X}_p)|\mathbf{f}$ is not Gaussian due to the $\mathbf{W}_{2,p}$ term in (5.10).

Pilot Design

The pilot design problem has the same formulation as (5.9) with $\mathbf{R}_{\Delta\mathbf{f}}$ given in (5.14). But for the case of estimated $\mathbf{G}$, the problem is even more difficult than that of the perfect $\mathbf{G}$ case. No analytical results have been reported. In [11], the designs for the perfect $\mathbf{G}$ case are used for simplicity. These design are expected to have good performance at least when the training power is high.

Estimation Error

The calculation of the power of the estimation error, i.e., MSE($\mathbf{f}$), is also more difficult for the case of estimated $\mathbf{G}$ than that of the perfect $\mathbf{G}$ case. In [9, 11], results for the special case of $M = 1$, i.e., the transmitter has a single antenna, was derived. In the derivation, it is assumed that P_s, P_r have the same order, i.e., P_s, $P_r \sim \mathcal{O}(P)$. The scaling order of MSE($\mathbf{f}$) with respect to the training power is derived.

Theorem 5.3. *[9, 11] For a relay network with single transmit antenna, R relay antennas, and N receive antennas, consider the estimation of* $\mathbf{f}$ *with estimated* $\mathbf{G}$ *in (5.13) and the aforementioned pilot design. When* $T_p = 1$, *MSE($\mathbf{f}$) =* $2R^2 \log_e P/P + \mathcal{O}(1/P)$; *when* $T_p \geq R$, *MSE($\mathbf{f}$) =* $\mathcal{O}(1/P)$.

This theorem shows similar error power behaviours as Theorem 5.2. For the case of $T_p = 1$, comparing the two theorems, we can see that for very high P, the MSE of the estimation with imperfect $\mathbf{G}$ doubles that of the estimation with perfect $\mathbf{G}$.

Similar to Sect. 5.1.3, results in Theorem 5.3 can be straightforwardly extended to obtain the following corollary for the general multiple-antenna multiple-relay network.

Corollary 5.2. *For a relay network with M transmit antennas, R relay antennas, and N receive antennas, if* $T_p \geq MR$, *MSE($\mathbf{f}$) =* $\mathcal{O}(1/P)$ *can be achieved using the estimation of* $\mathbf{f}$ *in (5.13) with estimated* $\mathbf{G}$ *and the aforementioned pilot design.*

5.1.5 Simulation Results on the MSE

In this subsection, some simulation results on the MSE of the TX-Relay channels $\mathbf{f}$ are shown.

The first simulation is on a network with $M = 1$, $R = 2$, and $N = 2$, i.e., a single antenna at the transmitter, a total of two antennas at the relays, and two antennas at the receiver. We set $P_s = RP_r$. For the training of the Relay-RX channel matrix $\mathbf{G}$, the MMSE estimation in (5.7) is used and $T_{p,\mathbf{G}} = 2 = R$. For the training of the TX-Relay channel vector $\mathbf{f}$, we consider two training time settings: $T_p = 1$ and $T_p = 2$ and three estimations: (1) the estimation in (5.8) with perfect $\mathbf{G}$, (2) the estimation in (5.8) with previously estimated $\mathbf{G}$, and (3) the LMMSE estimation in (5.13) with previously estimated $\mathbf{G}$. When $T_p = 1$, the total training time is

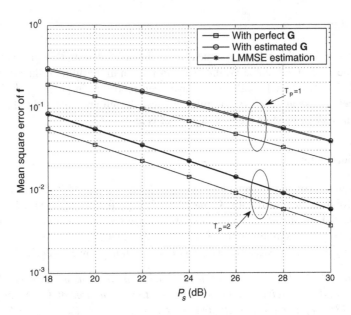

Fig. 5.1 MSEs of the TX-Relay channels for networks with $M = 1, R = 2, N = 2$

$T_{\text{training}} = T_{p,\mathbf{G}} + 2T_p = 4$ and $T_p < MR = 2$. When $T_p = 2$, the total training time is $T_{\text{training}} = T_{p,\mathbf{G}} + 2T_p = 6$ and $T_p = MR = 2$.

In Fig. 5.1, the MSEs on $\mathbf{f}$ are shown. It can be seen from this figure that for all cases, the MSE of $\mathbf{f}$ decreases as P increases. But when $T_p = 2$, for all three estimations, MSE($\mathbf{f}$) behaves as $\mathcal{O}(1/P)$; while when $T_p = 1$, for all three estimations, the decrease in MSE is slower. This observation is consistent with the results in Theorem 5.2 and 5.3. Comparing the estimation with perfect $\mathbf{G}$ and the LMMSE estimation with estimated $\mathbf{G}$, we see that the MSE of $\mathbf{f}$ with perfect $\mathbf{G}$ is about 3 dB smaller, i.e., the MSE of $\mathbf{f}$ with perfect $\mathbf{G}$ is about half of that with estimated $\mathbf{G}$. This again is consistent with the results in Theorem 5.2 and 5.3. Now we compare the estimation in (5.8) and the LMMSE estimation in (5.13) both with estimated $\mathbf{G}$. It can be seen that they have almost the same performance when the training power is high. For low training power, the LMMSE estimation, which takes into account the estimation error on $\mathbf{G}$, has slightly better performance. Comparing the MSEs of $T_p = 1$ with that $T_p = 2$, we see up to 8 dB improvement in the latter. This shows the importance of having enough training time.

The second simulated network has the following settings: $M = 2, R = 1$, and $N = 2$, i.e., two antennas at the transmitter, single relay antenna, and two antennas at the receiver. We again set $P_s = RP_r$. For the training of the Relay-RX channel matrix $\mathbf{G}$, 1 time slot is used, i.e., $T_{p,\mathbf{G}} = 1 = R$. The MMSE estimation in (5.7) is used. For the training of the TX-Relay channel vector $\mathbf{f}$, we again consider two training time settings: $T_p = 1$ and $T_p = 2$; and the same three estimations: (1) the estimation in (5.8) with perfect $\mathbf{G}$, (2) the estimation in (5.8) with previously estimated $\mathbf{G}$, and (3) the LMMSE estimation in (5.13) with previously estimated $\mathbf{G}$. When $T_p = 1$,

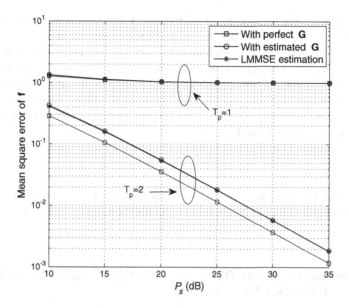

Fig. 5.2 MSEs of the TX-Relay channels for networks with $M = 2, R = 1, N = 2$

$T_p < MR = 2$ and the total training time is $T_{\text{training}} = T_{p,G} + 2T_p = 3$. When $T_p = 2$, $T_p = MR = 2$ and the total training time is $T_{\text{training}} = T_{p,G} + 2T_p = 5$. Figure 5.2 shows the MSEs. It can be seen from this figure that when $T_p = 1$, all three estimations have the same MSE and suffer an error floor. When $T_p = 2$, the MSEs of all three estimations behave as $\mathcal{O}(1/P)$. By comparing the estimation with perfect **G** and the LMMSE estimation with estimated **G**, for the case of $T_p = 2$, the MSE of the former is about 3 dB smaller. The estimation in (5.8) with estimated **G** and the LMMSE estimation in (5.13) with estimated **G** have the same performance. Comparing the MSEs of $T_p = 1$ with that $T_p = 2$, large improvement in MSE can be observed from the latter.

5.2 Training-Based DSTC

In this section, training-based DSTC, in other words, DSTC with estimated CSI, is studied. First, the training-based data-transmission model using DSTC is briefly explained, then several DSTC decodings under estimated CSI are represented. Finally, the performance of training-based DSTC and the latest diversity order results are demonstrated.

5.2.1 Training-Based Data-Transmission with DSTC

Consider the general multiple-antenna multiple-relay network shown in Fig. 3.1. The transmitter has M antennas, the receiver has N antennas, and the relays have R

antennas in total. The same as previous sections, $\mathbf{f}$ denotes the $MR \times 1$ TX-Relay channel vector and $\mathbf{G}$ denotes the $R \times N$ Relay-RX channel matrix.

Consider a block-fading model with coherence interval T_{total}. The channels are assumed to be unchanged during each coherence interval. Each coherence interval is split into two sub-intervals, where Sub-interval 1 is used for training, called the training interval; Sub-interval 2 is used for data-transmission, called the data-transmission interval. Denote the length of the training interval as T_{training} and the length of the data-transmission interval as T_{data}. Thus,

$$T_{\text{total}} = T_{\text{training}} + T_{\text{data}}.$$

We use the training schemes described in Sect. 5.1, where $T_{p,\mathbf{G}}$ time slots are used for the training of $\mathbf{G}$ and $2T_p$ time slots are used for the training of $\mathbf{f}$. Thus,

$$T_{\text{training}} = T_{p,\mathbf{G}} + 2T_p.$$

For the data-transmission, the DSTC scheme in Chap. 3 is used, where each of the two steps of DSTC takes T time slots. Thus,

$$T_{\text{data}} = 2T.$$

The relationship of these time-notation is represented in Fig. 5.3.

The same as previous section, $\hat{\mathbf{G}}$, $\hat{\mathbf{f}}$ denote estimations of the Relay-RX channel matrix and TX-relay channel vector obtained from the training process. $\Delta\mathbf{G}$ and $\Delta\mathbf{f}$ are the error matrix and error vector of the estimations of $\mathbf{G}$ and $\mathbf{f}$, respectively. Denote the transmit power of the transmitter as P_s and the transmit power of each relay antenna as P_r. For the date-transmission, denote the information matrix sent by the transmitter as $\mathbf{B}$, which satisfies $\mathbb{E}\{\text{tr}(\mathbf{B}^*\mathbf{B})\} = M$. $\mathbf{S}$ is the distributed space-time codeword formed at the receiver, defined in (3.8). To avoid the effect of the code design and focus on the diversity order results only, orthogonal distributed space-time codeword is used, i.e., $\mathbf{S}^*\mathbf{S} = \mathbf{I}_{MR}$. Note that different code designs only affect the coding gain of the network, not the diversity order, as long as fully diverse codes are used. The $T \times N$ matrix of the received signal during the data-transmission interval is denoted as $\mathbf{X}$. By using the results in Sect. 3.2, the transceiver equation of the data-transmission interval can be written as

$$\text{vec}(\mathbf{X}) = \sqrt{\beta}(\mathbf{G}^t \otimes \mathbf{I}_T)\tilde{\mathbf{S}}\mathbf{f} + \text{vec}(\mathbf{W}), \qquad (5.15)$$

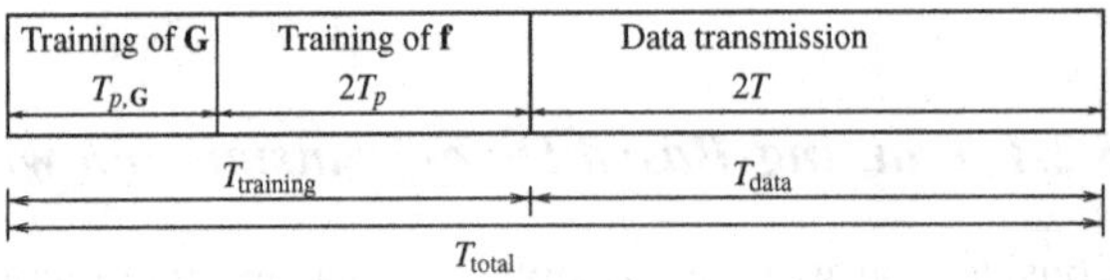

Fig. 5.3 A coherence interval for training-based DSTC

Training of G	Training of f	Data transmission
$T_{p,\mathbf{G}}$	$2T_p$	$2T$

T_{training}	T_{data}

T_{total}

where β is defined in (3.7) and $\tilde{\mathbf{S}} \triangleq \mathrm{diag}\{\mathbf{A}_1\mathbf{B}, \cdots, \mathbf{A}_R\mathbf{B}\}$. The transceiver equation can also be equivalently written as

$$\mathbf{X} = \sqrt{\beta}\mathbf{S}\mathbf{f}_{\mathrm{diag}}\mathbf{G} + \mathbf{W}, \qquad (5.16)$$

where

$$\mathbf{f}_{\mathrm{diag}} \triangleq \mathrm{diag}\{\mathbf{f}_1, \cdots, \mathbf{f}_R\},$$

with $\mathbf{f}_i$ the $M \times 1$ channel vector from the transmitter to the ith relay antenna. The equivalent noise term $\mathbf{W}$ is Gaussian whose covariance matrix is

$$\mathbf{R}_\mathbf{W} \triangleq \mathbf{I}_N + \alpha\mathbf{G}^*\mathbf{G},$$

where α is defined in (2.16).

5.2.2 Decodings of Training-Based DSTC

In this subsection, the decodings of training-based DSTC are explained, that is, the decoding schemes of DSTC with estimated CSI. First, the decoding of DSTC with perfect CSI is reviewed, then mismatched and matched decodings of DSTC with estimated CSI are studied.

Review on DSTC Decoding with Perfect CSI

With perfect CSI, the ML decoding of DSTC for the relay network is provided in Sect. 3.2 and copied here for the convenience of the presentation:

$$\mathrm{DEC}_0 : \underset{\mathbf{B}}{\arg\min}\, \mathrm{tr}\left\{\left(\mathbf{X} - \sqrt{\beta}\mathbf{S}\mathbf{f}_{\mathrm{diag}}\mathbf{G}\right)\mathbf{R}_\mathbf{W}^{-1}\left(\mathbf{X} - \sqrt{\beta}\mathbf{S}\mathbf{f}_{\mathrm{diag}}\mathbf{G}\right)^*\right\}. \qquad (5.17)$$

This decoding is named DEC_0.

It has been proved in Chap. 3 that when $T \geq MR$ and the CSI are known perfectly at the receiver, DEC_0 achieves full diversity order $\min\{M, N\}R$ when the transmit power is asymptotically high. The decoding complexity of DEC_0, however, is high even when real ODs are used. This is due to the noise covariance matrix $\mathbf{R}_\mathbf{W}$. Unlike the multiple-antenna system, the noise covariance matrix is not a multiple of the identity matrix, thus the decoding (5.17) cannot be decoupled into symbol-wise decoding. The joint decoding of all information symbols in $\mathbf{B}$ is required.

To reduce the complexity, a simplification of (5.17) can be obtained by omitting the noise covariance matrix:

$$\mathrm{DEC}_{0,\mathrm{simp}} : \underset{\mathbf{B}}{\arg\min}\, \left\|\mathbf{X} - \sqrt{\beta}\mathbf{S}\mathbf{f}_{\mathrm{diag}}\mathbf{G}\right\|_\mathrm{F}^2. \qquad (5.18)$$

This is equivalent to ignoring the correlation and the difference in variances among the equivalent noise entries. Another way to understand this suboptimal decoding is as follows. If $\mathbf{G}^*\mathbf{G}$ is replaced with its expectation $R\mathbf{I}_R$ in $\mathbf{R_W}$, it can be obtained that $\mathbf{R_W} \approx (1 + \alpha R)\mathbf{I}_R$. This approximation becomes equality with probability 1 as $R \to \infty$. Thus the simplified decoding $\text{DEC}_{0,\text{simp}}$ approaches the ML decoding DEC_0 when R increases to infinity. While considering the orthogonal structure of $\mathbf{S}$, the simplified decoding in (5.18) can be decoupled into symbol-wise decoding, i.e., separately decoding of each information symbol. The decoding complexity is largely reduced. For further details on how the decoding can be performed symbol-by-symbol, please refer to [4, 11, 12]. Simulation shows that $\text{DEC}_{0,\text{simp}}$ performs almost the same as the ML decoding, DEC_0.

Training-Based Mismatched Decoding

In reality, perfect CSI at the receiver is impossible and only erroneous channel estimations are available. Now we work on the training-based system equation. Replacing $\mathbf{G}$ and $\mathbf{f}$ in the transceiver equation by $\hat{\mathbf{G}} + \Delta\mathbf{G}$ and $\hat{\mathbf{f}} + \Delta\mathbf{f}$, respectively, the following training-based data-transmission equation can be obtained:

$$\text{vec}(\mathbf{X}) = \sqrt{\beta}\left(\hat{\mathbf{G}}^t \otimes \mathbf{I}_T\right)\tilde{\mathbf{S}}\hat{\mathbf{f}} + \mathbf{w}_e, \tag{5.19}$$

where $\mathbf{w}_e$ is the noise-plus-estimation-error term. It is defined as

$$\mathbf{w}_e \triangleq \sqrt{\beta}\left[\left(\hat{\mathbf{G}}^t \otimes \mathbf{I}_T\right)\tilde{\mathbf{S}}\Delta\mathbf{f} + \left(\Delta\mathbf{G}^t \otimes \mathbf{I}_T\right)\tilde{\mathbf{S}}\hat{\mathbf{f}} + \left(\Delta\mathbf{G}^t \otimes \mathbf{I}_T\right)\tilde{\mathbf{S}}\Delta\mathbf{f}\right]$$
$$+ \text{vec}(\mathbf{W}_1 + \mathbf{W}_2 + \mathbf{W}), \tag{5.20}$$

where

$$\mathbf{W}_1 \triangleq \sqrt{\alpha}\,[\mathbf{A}_1\mathbf{v}_1 \ \cdots \ \mathbf{A}_R\mathbf{v}_R]\,\hat{\mathbf{G}}$$

and

$$\mathbf{W}_2 \triangleq \sqrt{\alpha}\,[\mathbf{A}_1\mathbf{v}_1 \ \cdots \ \mathbf{A}_R\mathbf{v}_R]\,\Delta\mathbf{G}.$$

Given the estimations $\hat{\mathbf{G}}$ and $\hat{\mathbf{f}}$ at the receiver, the most straightforward decoding is to ignore the estimation error, by which the transmission equation is seen as

$$\text{vec}(\mathbf{X}) = \sqrt{\beta}\left(\hat{\mathbf{G}}^t \otimes \mathbf{I}_T\right)\tilde{\mathbf{S}}\hat{\mathbf{f}} + \text{vec}(\mathbf{W}),$$

or equivalently,

$$\mathbf{X} = \sqrt{\beta}\mathbf{S}\hat{\mathbf{f}}_{\text{diag}}\hat{\mathbf{G}} + \mathbf{W},$$

where $\hat{\mathbf{f}}_{\text{diag}} \triangleq \text{diag}\{\hat{\mathbf{f}}_1, \cdots, \hat{\mathbf{f}}_R\}$. The ML decoding for this case is thus

$$\text{DEC}_1 \,:\, \arg\min_{\mathbf{B}} \text{tr}\left\{\left(\mathbf{X} - \sqrt{\beta}\mathbf{S}\hat{\mathbf{f}}_{\text{diag}}\hat{\mathbf{G}}\right)\hat{\mathbf{R}}_{\mathbf{W}}^{-1}\left(\mathbf{X} - \sqrt{\beta}\mathbf{S}\hat{\mathbf{f}}_{\text{diag}}\hat{\mathbf{G}}\right)^*\right\}, \tag{5.21}$$

where

$$\hat{\mathbf{R}}_{\mathbf{W}} \triangleq \mathbf{I}_N + \alpha \hat{\mathbf{G}}^* \hat{\mathbf{G}}.$$

In obtaining DEC_1, the estimated channels are seen as perfect and the estimation errors are ignored. The DEC_1 formula thus does not match the training-based data-transmission equation. It is called *mismatched decoding*.

Similar to the perfect CSI case, by approximating $\hat{\mathbf{G}}\hat{\mathbf{G}}^*$ with its expectation $\frac{RP_rT_{p,G}}{1+P_rT_{p,G}}\mathbf{I}_R$ in $\hat{\mathbf{R}}_{\mathbf{W}}$, a simplified mismatched decoding $DEC_{1,\text{simp}}$ is obtained:

$$DEC_{1,\text{simp}} : \arg\min_{\mathbf{B}} \left\| \mathbf{X} - \sqrt{\beta}\mathbf{S}\left(\hat{\mathbf{f}}_{\text{diag}}\hat{\mathbf{G}}\right) \right\|_F^2 . \tag{5.22}$$

When the code-matrix $\mathbf{S}$ has OD structure, the decoding $DEC_{1,\text{simp}}$ can be performed symbol-by-symbol and has low complexity.

Training-Based Matched Decoding

In this section, a matched decoding, where the channel estimation errors are taken into account, is studied. We use the estimations of $\hat{\mathbf{G}}$ and $\hat{\mathbf{f}}$ in (5.7) and (5.13). From the definition of the noise-plus-estimation-error term $\mathbf{w}_e$ in the training-based transceiver equation (5.19), it can be seen that $\mathbf{w}_e$ is not Gaussian, which makes further analysis intractable. In what follows, an approximation of the training-based transceiver equation is considered. We ignore the terms $\sqrt{\beta}(\Delta\mathbf{G}^t \otimes \mathbf{I}_T)\tilde{\mathbf{S}}\hat{\mathbf{f}}$, $\beta(\Delta\mathbf{G}^t \otimes \mathbf{I}_T)\tilde{\mathbf{S}}\Delta\mathbf{f}$, $\mathbf{W}_1$, and $\mathbf{W}_2$ to approximate $\mathbf{w}_e$ as

$$\mathbf{w}_e \approx \mathbf{w}_e' \triangleq \sqrt{\beta}\left(\hat{\mathbf{G}}^t \otimes \mathbf{I}_T\right)\tilde{\mathbf{S}}\Delta\mathbf{f} + \text{vec}(\mathbf{W}).$$

To explain this approximation, we study the power of each of the noise or estimation error term in (5.20). From the definition of β and α,

$$\beta = \frac{T}{R}P + \mathcal{O}(1) \text{ and } \alpha = \frac{1}{R} + \mathcal{O}\left(\frac{1}{P}\right).$$

When $T_{p,G} \geq R$, it has been shown that $\text{MSE}(\mathbf{G}) = \mathcal{O}(1/P)$. Also, $\text{MSE}(\mathbf{f}) = \mathcal{O}(1/P)$ when $T_p \geq MR$. But when $T_p \leq MR$, it may happen that $\text{MSE}(\mathbf{f}) = \mathcal{O}(\log_e P/P)$. Thus, first of all, the power of the term $\beta(\Delta\mathbf{G}^t \otimes \mathbf{I}_T)\tilde{\mathbf{S}}\Delta\mathbf{f}$ is much smaller compared with those of other noise terms when the transmit power is high. Also, the powers of $\sqrt{\beta}(\Delta\mathbf{G}^t \otimes \mathbf{I}_T)\tilde{\mathbf{S}}\hat{\mathbf{f}}$, $\mathbf{W}_1$, and $\mathbf{W}_2$ scale as $\mathcal{O}(1)$. But the power of $\sqrt{\beta}(\hat{\mathbf{G}}^t \otimes \mathbf{I}_T)\tilde{\mathbf{S}}\Delta\mathbf{f}$ can scale as $\log_e P$, which is much larger than other terms. This term is kept in the transceiver equation. The last term $\mathbf{W}$ is also kept because its deletion may result in singularity or ill-condition problem, when the inverse of the noise covariance matrix is conducted in the decoding.

With the aforementioned approximation on the noise-plus-estimation-error term, an approximate transceiver equation is derived as

$$\text{vec}(\mathbf{X}) = \sqrt{\beta} \left(\hat{\mathbf{G}}^t \otimes \mathbf{I}_T \right) \tilde{\mathbf{S}} \hat{\mathbf{f}} + \mathbf{w}'_e. \tag{5.23}$$

It can be shown straightforwardly that $\mathbf{w}'_e$ has zero-mean. Its covariance matrix can be calculated to be

$$\mathbf{R}_{\mathbf{w}'_e} = \beta \left[\left(\hat{\mathbf{G}}^t \otimes \mathbf{I}_T \right) \tilde{\mathbf{S}} \right] \mathbf{R}_{\Delta \mathbf{f}} \left[\left(\hat{\mathbf{G}}^t \otimes \mathbf{I}_T \right) \tilde{\mathbf{S}} \right]^* + \mathbf{I}_{TN}, \tag{5.24}$$

where $\mathbf{R}_{\Delta \mathbf{f}}$ is the covariance matrix of $\Delta \mathbf{f}$ provided in (5.14).

If $\Delta \mathbf{f}$ is Gaussian, since $\mathbf{W}$ is Gaussian, for given channel estimations $\hat{\mathbf{G}}$ and $\hat{\mathbf{f}}$, the approximate noise term $\mathbf{w}'_e$ is a Gaussian vector. The conditional probability density function (PDF) of the receive signal matrix $\text{vec}(\mathbf{X})|\hat{\mathbf{G}}, \hat{\mathbf{f}}, \mathbf{B}$ is

$$p(\text{vec}(\mathbf{X})|\hat{\mathbf{G}}, \hat{\mathbf{f}}, \mathbf{B}) = \frac{1}{\pi^{TR} \det\left(\mathbf{R}_{\mathbf{w}'_e}\right)} e^{-\left(\text{vec}(\mathbf{X}) - \sqrt{\beta}\left[\left(\hat{\mathbf{G}}^t \otimes \mathbf{I}_T\right)\tilde{\mathbf{S}}\right]\hat{\mathbf{f}}\right)^* \mathbf{R}_{\mathbf{w}'_e}^{-1} \left(\text{vec}(\mathbf{X}) - \beta\left[\left(\hat{\mathbf{G}}^t \otimes \mathbf{I}_T\right)\tilde{\mathbf{S}}\right]\hat{\mathbf{f}}\right)}.$$

By maximizing this PDF, the following decoding rule is obtained:

$$\text{DEC}_2 : \arg\min_{\mathbf{B}} \left[\ln \det\left(\mathbf{R}_{\mathbf{w}'_e}\right) + \right.$$
$$\left. \left(\text{vec}(\mathbf{X}) - \sqrt{\beta}\left[\left(\hat{\mathbf{G}}^t \otimes \mathbf{I}_T\right)\tilde{\mathbf{S}}\right]\hat{\mathbf{f}}\right)^* \mathbf{R}_{\mathbf{w}'_e}^{-1} \left(\text{vec}(\mathbf{X}) - \beta\left[\left(\hat{\mathbf{G}}^t \otimes \mathbf{I}_T\right)\tilde{\mathbf{S}}\right]\hat{\mathbf{f}}\right) \right]. \tag{5.25}$$

Note that $\mathbf{R}_{\mathbf{w}'_e}$ depends on the information matrix $\mathbf{B}$ via $\tilde{\mathbf{S}}$. Since in the derivation of this decoding, $\Delta \mathbf{f}$ is treated as Gaussian and some lower order noise and estimation error terms are neglected, this decoding is not the optimal ML decoding of the training-based DSTC transmission. But it approaches the optimal ML decoding when the training power increases. The exact ML decoding will perform no worse than DEC_2.

In DEC_2, the covariance matrix of the estimation error term is incorporated through $\mathbf{R}_{\mathbf{w}'_e}$. This decoding is called *matched decoding*, because as opposed to DEC_1, it takes into account the estimation error and matches the approximate training-based data-transmission equation. Simulation shows that DEC_2 achieves better performance than the mismatched decoding DEC_1, when the training interval is short. It sometimes achieves a higher diversity order.

However, this performance improvement of the matched decoding comes with a price on computational load. The prominent difference of the matched decoding to the mismatched one is in the covariance matrix $\mathbf{R}_{\mathbf{w}'_e}$, which incorporates the channel estimation error and also depends on the information matrix $\mathbf{B}$. DEC_2 thus cannot be reduced to decoupled symbol-wise decoding. For detailed analysis on the computational loads of the mismatched and matched decodings, please refer to [11].

5.2.3 Performance of Training-Based DSTC

In this subsection, performance of training-based DSTC under different decoding rules is discussed. We focus on recent diversity order results.

The first to consider is the mismatched decoding in (5.21), where the estimated CSI is treated as perfect. It has been proved in Chaps. 2 and 3 that with perfect CSI, DSTC achieves full diversity order when the transmit power is asymptotically high. By comparing the transceiver equation of the training-based DSTC given in (5.19) and (5.20) with that of DSTC with perfect CSI in (3.6), it can be seen intuitively that if the power of the equivalent noise-plus-estimation-error term in (5.20) has the same scaling as that of the equivalent noise term in the perfect CSI case, training-based DSTC with mismatched decoding should also achieve full diversity order. The equivalent noise term for perfect CSI case scales as $\mathcal{O}(1)$. Thus, when training-based DSTC with mismatched decoding is employed, for full diversity order, MSE($\mathbf{G}$) and MSE($\mathbf{f}$) should scale as $\mathcal{O}(1/P)$.

It is mentioned in Sect. 5.1.2 that MSE($\mathbf{G}$) scales as $\mathcal{O}(1/P)$ when $T_{p,\mathbf{G}} \geq R$. In Theorem 5.2 and 5.3, we mention that MSE($\mathbf{f}$) scales as $\mathcal{O}(1/P)$ when $T_p \geq MR$. Thus, if the total training time is no less than $(2M+1)R$, full diversity order should be obtained with mismatched decoding. This result was proved in [10] for perfect $\mathbf{G}$ case.

When $T_p < MR$, we cannot guarantee the scaling of $\mathcal{O}(1/P)$ for MSE($\mathbf{f}$). For the special case of $M = 1$, Theorem 5.2 and 5.3 show that if $T_p = 1$, which is the shortest training time, MSE($\mathbf{f}$) scales as $\mathcal{O}(\log_e P/P)$. This leads to diversity loss for networks with multiple relay and receive antennas [11]. Another result on diversity order has been proved in [10], which says that when $R = 1$ or $N = 1$, for any $T_p \leq MR$, full diversity order cannot be achieved. The diversity order of multiple-antenna multiple-relay network with a general training time using mismatched decoding is still unknown.

Next, we consider matched decoding given in (5.25), since channel estimation errors are taken into account, it is expected to perform better than mismatched decoding. In [11], it has been shown by simulation that for some network scenarios and training time settings (e.g., $M = 1$ and $T_p = 1$) where mismatched decoding cannot achieve full diversity order, matched decoding may achieve a higher diversity order and sometimes full diversity order. But there is no analytic results on the achievable diversity order for matched decoding so far.

In the remaining part of this section, numerical simulation results on the block error rates of training-based DSTC are shown.

The first simulation is on the block error rate of training-based DSTC in a network with $M = 1$, $R = 2$, $N = 2$ and $P_s = RP_r$. For the training of the Relay-RX channel matrix $\mathbf{G}$, the MMSE estimation in (5.7) is used where $T_{p,\mathbf{G}} = 2 = R$. For the training of the TX-Relay channel vector $\mathbf{f}$, we use the LMMSE estimation in (5.13) with two training time settings: $T_p = 1$ and $T_p = 2$. Note that when $T_p = 1$, $T_p < MR = 2$ and the total training time is $T_{\text{training}} = T_{p,\mathbf{G}} + 2T_p = 4$. When $T_p = 2$, $T_p = MR = 2$ and the total training time is $T_{\text{training}} = T_{p,\mathbf{G}} + 2T_p = 6$. Three decodings are considered for the data-transmission interval. The first decoding

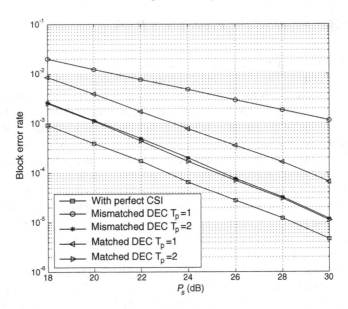

Fig. 5.4 Block error rates for a network with $M = 1$, $R = 2$, $N = 2$ using training-based DSTC

is DEC_0, which is the decoding with perfect CSI. It is used as a benchmark. The second decoding is DEC_1, the mismatched decoding in (5.21). The third decoding is DEC_2, the matched decoding in (5.25). It can be seen from Fig. 5.4 that when $T_p = MR = 2$, both the mismatched decoding and the matched decoding achieve full diversity order, which is 2. The matched decoding has slightly better performance. Compared with the perfect CSI case, the matched decoding with estimated channels is about 2 dB worse. When $T_p = 1 < MR = 2$, however, the mismatched decoding loses diversity order. It only achieves diversity order 1. But the matched decoding still achieves full diversity order, which is 2. When T_p increases from 1 to 2, the network performance increases by approximately 5 dB for the matched decoding.

The next simulation is on the block error rate of training-based DSTC in a network with $M = 2$, $R = 1$, $N = 2$ and $P_s = RP_r$. For the training of the Relay-RX channel matrix $\mathbf{G}$, the MMSE estimation in (5.7) is used where $T_{p,\mathbf{G}} = 1 = R$. For the training of the TX-Relay channel vector $\mathbf{f}$, we use the LMMSE estimation in (5.13) and again consider two training time settings: $T_p = 1$ and $T_p = 2$. Note that when $T_p = 1$, $T_p < MR = 2$ and the total training time is $T_{\text{training}} = T_{p,\mathbf{G}} + 2T_p = 3$. When $T_p = 2$, $T_p = MR = 2$ and the total training time is $T_{\text{training}} = T_{p,\mathbf{G}} + 2T_p = 5$. The same three decodings are considered for the data-transmission interval. It can be seen from Fig. 5.5 that when $T_p = MR = 2$, both the mismatched decoding and the matched decoding achieve full diversity order, which is 2. The matched decoding has slightly better performance than the mismatched one. Compared with the perfect CSI case, the matched decoding with estimated channels is about 3 dB worse. When $T_p = 1 < MR = 2$, however, the mismatched decoding loses all diversity. Its

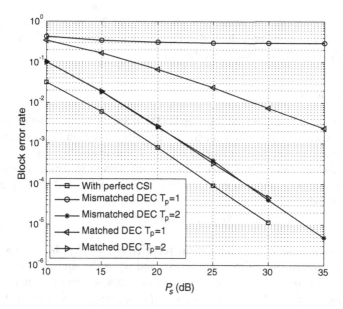

Fig. 5.5 Block error rates for a network with $M = 2$, $R = 1$, $N = 2$ using training-based DSTC

diversity order is 0. The diversity order of the matched decoding is 1. So matched decoding achieves a higher diversity order, although it does not achieve full diversity order. When T_p increases from 1 to 2, for both mismatched and matched decodings, the network performance is improved largely.

5.3 End-to-End Channel Estimation for Single-Antenna Multiple-Relay Network

Section 5.1 is on the individual estimate of both the Relay-RX channel matrix $\mathbf{G}$ and the TX-Relay channel vector $\mathbf{f}$ at the receiver. When $\mathbf{G}$ and $\mathbf{f}$ are estimated, the DSTC (mismatched and matched) decoding can be conducted using (5.17) and (5.25). Another equivalent (mismatched) decoding formula for DSTC is in (2.51), where an estimation on $\mathbf{G}$ (to calculate $\mathbf{R_W}$) and an estimation on $\mathbf{H}$ is required. One can of course obtain an estimation of $\mathbf{H}$ from estimations on $\mathbf{G}$ and $\mathbf{f}$. But another possibility is to estimate $\mathbf{H}$ directly. $\mathbf{H}$ contains the effective channels from all transmit antennas to all receive antennas via the relay antennas. It is called the *end-to-end channel matrix*. This section is on the end-to-end channel estimation for single-antenna multiple-relay network. The next section considers the multiple-antenna case. All channels are assumed to be i.i.d. circularly symmetric complex Gaussian with zero-mean and unit-variance. The magnitude of each channel coefficient follows Rayleigh distribution. The channels are also assumed to remain constant during training.

For a single-antenna relay network shown in Fig. 2.2, the end-to-end channel matrix reduces to a vector $\mathbf{h}$ defined in (2.17). In what follows, we talk about the training scheme and the estimation rule for $\mathbf{h}$.

Training Scheme

The training scheme follows the two-step DSTC in Sect. 2.2, where in the first step, a $T_p \times 1$ pilot vector $\mathbf{b}_p$ is sent from the transmitter to the relays, the relays then linearly transform the signal vectors they receive via $T_p \times T_p$ unitary matrices $\mathbf{A}_{1,p}, \cdots, \mathbf{A}_{R,p}$ and forward to the receiver in the second step. Each training step takes T_p times slots and the total training duration is $2T_p$. The following end-to-end training equation can be obtained:

$$\mathbf{x}_p = \sqrt{\beta_p}\mathbf{S}_p\mathbf{h} + \mathbf{w}_p,$$

where $\mathbf{x}_p$ is the observation vector at the receiver, $\mathbf{S}_p \triangleq \begin{bmatrix} \mathbf{A}_{1,p}\mathbf{b}_p & \cdots & \mathbf{A}_{R,p}\mathbf{b}_p \end{bmatrix}$ is the pilot distributed space-time code-matrix, and $\mathbf{w}_p$ is the equivalent noise. For a given realization of $\mathbf{g}$, $\mathbf{w}_p$ is a Gaussian random vector, where its mean is zero and its covariance matrix is $\mathbf{R}_{\mathbf{w}_p} \triangleq (1 + \alpha_p\|\mathbf{g}\|_F^2)\mathbf{I}_{T_p}$. α_p and β_p are defined in (5.3).

End-to-End Channel Estimation Rule

The end-to-end channel estimation problem is to estimate $\mathbf{h}$ based on the observation $\mathbf{x}_p$. Both the maximum-likelihood (ML) estimation and the LMMSE estimation are considered in this section.

For a given channel realization, we have the following conditional PDF:

$$p(\mathbf{x}_p|\mathbf{f}, \mathbf{g}) = \frac{1}{(2\pi)^R(1 + \alpha_p\|\mathbf{g}\|_F^2)^R}e^{-(1+\alpha_p\|\mathbf{g}\|_F^2)^{-1}\|\mathbf{x}_p - \sqrt{\beta_p}\mathbf{S}_p\mathbf{h}\|_2^2}.$$

Thus,

$$p(\mathbf{x}_p|\mathbf{h}) = \int p(\mathbf{x}_p|\mathbf{h}, \mathbf{g})\,p(\mathbf{g})d\mathbf{g} = \int p(\mathbf{x}_p|\mathbf{f}, \mathbf{g})p(\mathbf{g})d\mathbf{g}$$

$$= \int \frac{1}{(2\pi)^R(1 + \alpha_p\|\mathbf{g}\|_F^2)^R}e^{-(1+\alpha_p\|\mathbf{g}\|_F^2)^{-1}\|\mathbf{x}_p - \sqrt{\beta_p}\mathbf{S}_p\mathbf{h}\|_2^2}\frac{1}{(2\pi)^R}e^{-\|\mathbf{g}\|_2^2}d\mathbf{g}.$$

Note that although with the relationship $\mathbf{h} = \mathbf{f} \circ \mathbf{g}$, for a given $\mathbf{g}$, the possible values of $\mathbf{h}$ is not constrained by the given value of $\mathbf{g}$. Thus the ML estimation of $\mathbf{h}$ can be derived as follows:

$$\hat{\mathbf{h}}_{\text{ML}} = \arg\max_{\mathbf{h}} p(\mathbf{x}_p|\mathbf{h}) = \arg\max_{\mathbf{h}} \left\| \mathbf{x}_p - \sqrt{\beta}\mathbf{S}_p\mathbf{h} \right\|_2^2 = \frac{1}{\sqrt{\beta}}(\mathbf{S}_p^*\mathbf{S}_p)^{-1}\mathbf{S}_p^*\mathbf{x}_p.$$

The above ML estimation requires that $T_p \geq R$ for $(\mathbf{S}_p^* \mathbf{S}_p)^{-1}$ to exist. This ML estimation result happens to be the same as that of a multiple-antenna system with multiple transmit antennas and single receive antenna [2]. But the derivations are different. This ML estimation is also the same as the least square (LS) estimation in [1].

The LMMSE estimation can be derived using Bayesian Gauss-Markov theorem [1, 6] to be

$$\hat{\mathbf{h}}_{\text{LMMSE}} = \sqrt{\beta_p} \left[(1 + \alpha_p) \mathbf{I}_M + \beta_p \mathbf{S}_p^* \mathbf{S}_p \right]^{-1} \mathbf{S}_p^* \mathbf{x}_p.$$

5.4 End-to-End Channel Estimation for Multiple-Antenna Multiple-Relay Network

This section considers the estimation of the end-to-end channels in a general multiple-antenna multiple-relay network. The network diagram is illustrated in Sect. 3.2, where the transmitter has M antennas, the receiver has N antennas, and there are in total R antennas at the relays. The end-to-end channel estimation problem is to estimate all channels from each antenna of the transmitter to each antenna of the receiver via each relay antenna. There are in total MNR channel coefficients to be estimated at the receiver. In this book, only independent wireless channels are considered, thus, channels corresponding to one relay antenna are independent to those corresponding to another relay antenna. So the channel estimation can be considered separately and sequentially. Thus, in what follows, we study the estimation of the end-to-end channels corresponding one relay antenna. Estimations of channels corresponding to other relay antennas are the same.

Consider the network shown in Fig. 5.6, where there are M transmit antennas, N receive antennas, and one relay antenna. Let

$$\mathbf{f} \triangleq \begin{bmatrix} f_1 & f_2 & \cdots & f_M \end{bmatrix}^t$$

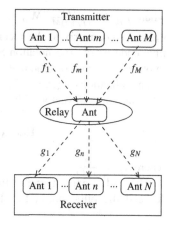

Fig. 5.6 Wireless relay network with multiple transmit antennas, multiple receive antennas, and single relay antenna

be the $M \times 1$ channel vector from the transmitter to the relay, where f_m is the channel from the mth transmit antenna to the relay antenna. Let

$$\mathbf{g} \triangleq \begin{bmatrix} g_1 & g_2 & \cdots & g_N \end{bmatrix}$$

be the $1 \times N$ channel vector from the relay to the receiver, where g_n is the channel from the relay antenna to the nth receive antenna. The end-to-end channel from the mth transmit antenna to the nth receive antenna via the relay antenna is $f_m g_n$. The $M \times N$ end-to-end channel matrix of the network is

$$\mathbf{H} \triangleq \begin{bmatrix} f_1 g_1 & f_1 g_2 & \cdots & f_1 g_N \\ f_2 g_1 & f_2 g_2 & \cdots & f_2 g_N \\ \vdots & \vdots & \ddots & \vdots \\ f_M g_1 & f_M g_2 & \cdots & f_M g_N \end{bmatrix} = \mathbf{fg}. \tag{5.26}$$

5.4.1 Training Scheme

The training process takes two steps. During the first step, the transmitter sends $\sqrt{P_s T_p / M} \mathbf{B}_p$, where the $T_p \times M$ matrix $\mathbf{B}_p$ is the pilot, normalized as $\mathrm{tr}\{\mathbf{B}_p^* \mathbf{B}_p\} = M$. This normalization implies that the average transmit power of the transmitter is P_s per transmission. The relay then amplifies the signals it receives and forwards to the receiver in the second step. The fixed gain relay power coefficient $\sqrt{P_r / (P_s + 1)}$ is used [7, 8], so the average relay transmit power is P_r. This relay processing is a special case of DSTC for networks with single relay antenna explain in Chap. 3. By using the transceiver equation in (3.6) of Chap. 3 for the special case of $R = 1$ and $\mathbf{A}_1 = \mathbf{I}_{T_p}$, the transceiver equation of the training interval is

$$\mathbf{X}_p = \sqrt{\beta_p} \mathbf{S}_p \mathbf{H} + \mathbf{W}_p, \tag{5.27}$$

where $\beta_p \triangleq \frac{P_s P_r T_p}{(P_s + 1) M}$. $\mathbf{W}_p$ is the equivalent noise matrix during training, $\mathbf{X}_p$ is the received matrix at the receiver, and $\mathbf{S}_p \triangleq \mathbf{B}_p$ is the pilot distributed space-time code-word. The notation $\mathbf{S}_p$ is used to be consistent with the representation in other chapters. Assume that all noise entries are i.i.d. following $\mathcal{CN}(0, 1)$. For a given realization of $\mathbf{g}$, $\mathrm{vec}(\mathbf{W}_p)$ can be shown to be a circularly symmetric complex Gaussian random vector with zero mean, and its covariance matrix is

$$\mathbf{R}_{\mathrm{vec}(\mathbf{W}_p)} \triangleq \left(\mathbf{I}_N + \alpha_p \mathbf{g}^t \bar{\mathbf{g}} \right) \otimes \mathbf{I}_T, \tag{5.28}$$

where $\alpha_p \triangleq \frac{P_r}{P_s + 1}$.

The end-to-end channel estimation is to estimate $\mathbf{H}$ based on the observation $\mathbf{X}_p$ and the pilot matrix $\mathbf{S}_p$. The training equation (5.27) has the same format as the traditional Gaussian observation model [6] or the training equation for a multiple-antenna system without relaying [2]. However, it differs to the traditional Gaussian model in two aspects. First, in a traditional Gaussian model, the noises are independent of the parameters to be estimated. But the covariance matrix of the noise term in (5.27) is a function of $\mathbf{g}$, which is related to $\mathbf{H}$, the matrix to be estimated. This is due to the propagation of the relay noise. Second, in the traditional Gaussian model, entries of the vector/matrix to be estimated are independent. Here, for the relay network, the $M \times N$ end-to-end channel matrix $\mathbf{H}$ has rank-1. Thus, entries of $\mathbf{H}$ are related. Only $M + N - 1$ of the total MN entries in $\mathbf{H}$ are independent.

5.4.2 Entry-Based Estimation

From (5.27), if entries of $\mathbf{H}$ are regarded as independent and the relation between $\mathbf{H}$ and the noise covariance matrix is ignored, the estimation problem can be seen as a traditional Gaussian one. The ML estimation and the LMMSE estimation can be derived straightforwardly as [5, 6]

$$\hat{\mathbf{H}}_{\text{entry,ML}} = \frac{1}{\sqrt{\beta_p}}(\mathbf{S}_p^*\mathbf{S}_p)^{-1}\mathbf{S}_p^*\mathbf{X}_p, \tag{5.29}$$

$$\hat{\mathbf{H}}_{\text{entry,LMMSE}} = \sqrt{\beta_p}\left[(1+\alpha_p)\mathbf{I}_M + \beta_p\mathbf{S}_p^*\mathbf{S}_p\right]^{-1}\mathbf{S}_p^*\mathbf{X}_p. \tag{5.30}$$

These are straightforward extensions of the estimation schemes in Sect. 5.3 to multiple-antenna relay network. They are called *entry-based estimations* because they directly estimate each entry of the matrix $\mathbf{H}$ without considering the connections among them. With probability 1, the channel estimation results in (5.29) and (5.30) are not rank-1.

5.4.3 SVD-Based Estimation

Entry-based estimations are straightforward. But they do not consider the special structure of the channel matrix, thus they may not be optimal. An SVD-based ML estimation was proposed in [5], which takes into account the rank-1 structure of $\mathbf{H}$.

SVD of the End-to-End Channel Matrix

By considering the rank-1 structure, the channel matrix $\mathbf{H}$ can be decomposed as

$$\mathbf{H} = a\tilde{\mathbf{f}}\tilde{\mathbf{g}}, \tag{5.31}$$

where

$$a \triangleq \|\mathbf{f}\|_F \|\mathbf{g}\|_F, \quad \tilde{\mathbf{f}} \triangleq \frac{\mathbf{f}}{\|\mathbf{f}\|_F}, \quad \text{and} \quad \tilde{\mathbf{g}} \triangleq \frac{\mathbf{g}}{\|\mathbf{g}\|_F}. \tag{5.32}$$

$\tilde{\mathbf{f}}$ and $\tilde{\mathbf{g}}$ are the directions of $\mathbf{f}$ and $\mathbf{g}$, respectively. They have unit-norm. Since entries of $\mathbf{f}$ and $\mathbf{g}$ are i.i.d. $\mathcal{CN}(0, 1)$, a, $\tilde{\mathbf{f}}$, $\tilde{\mathbf{g}}$ are mutually independent. The decomposition in (5.31) is actually the SVD of $\mathbf{H}$, where a is the only non-zero singular value, and $\tilde{\mathbf{f}}$ and $\tilde{\mathbf{g}}$ are the corresponding left singular vector and the Hermitian of the right singular vector. The decomposition in (5.31) provides a map between $\mathbf{H}$ and the 3-tuple $\left(a, \tilde{\mathbf{f}}, \tilde{\mathbf{g}}\right)$. The estimation of $\mathbf{H}$ can thus be transformed to the estimation of $\left(a, \tilde{\mathbf{f}}, \tilde{\mathbf{g}}\right)$. The estimation based on this decomposition is called *SVD-based estimation*.

Note that the total number of real dimensions in the 3-tuple $\left(a, \tilde{\mathbf{f}}, \tilde{\mathbf{g}}\right)$ is $1+(2M-1)$ $+(2N-1) = 2(M+N)-1$. The total number of real dimensions in the rank-1 matrix $\mathbf{H}$ is $2(M+N-1)$, which is 1 dimension less. This is because that SVD decomposition is not unique. For any angle $\theta \in [0, 2\pi)$, from (5.32), $\mathbf{H} = a\left(e^{j\theta}\tilde{\mathbf{f}}\right)\left(e^{-j\theta}\tilde{\mathbf{g}}\right)$. For two different 3-tuples of estimations, $\left(\hat{a}, \hat{\tilde{\mathbf{f}}}, \hat{\tilde{\mathbf{g}}}\right)$ and $\left(\hat{a}, e^{j\theta}\hat{\tilde{\mathbf{f}}}, e^{-j\theta}\hat{\tilde{\mathbf{g}}}\right)$, the same estimation on $\mathbf{H}$ will be obtained. The map from $\mathbf{H}$ to $\left(a, \tilde{\mathbf{f}}, \tilde{\mathbf{g}}\right)$ is a set-valued map.

SVD-Based ML Estimation Results

Now we consider the SVD-based estimation of $\mathbf{H}$ under the ML estimation framework. The ML estimation maximizes the likelihood function, which is the conditional PDF of the observation [6]. That is $\left(\hat{a}, \hat{\tilde{\mathbf{f}}}, \hat{\tilde{\mathbf{g}}}\right) = \arg\max_{(a,\tilde{\mathbf{f}},\tilde{\mathbf{g}})} p(\mathbf{X}_p|a, \tilde{\mathbf{f}}, \tilde{\mathbf{g}})$. Due to the dependence of $\mathbf{R}_{\text{vec}(\mathbf{W}_p)}$ on $\mathbf{g}$ and the relationship of a with $\|\mathbf{g}\|_F$, $\|\mathbf{f}\|_F$, the exact ML estimation is difficult to derive. In the following, the ML channel estimation under the ideal assumption that $\|\mathbf{g}\|_F$ is known is derived first. Then an estimator for the practical case that $\|\mathbf{g}\|_F$ is unknown is obtained [5].

From the training equation (5.27), we have

$$p(\mathbf{X}_p|\mathbf{f}, \mathbf{g}) = \frac{1}{(2\pi)^{T_pN} \det \mathbf{R}_{\text{vec}(\mathbf{W}_p)}} e^{-\text{vec}\left(\mathbf{X}_p - \sqrt{\beta_p}\mathbf{S}_p\mathbf{H}\right)^* \mathbf{R}_{\text{vec}(\mathbf{W}_p)}^{-1} \text{vec}\left(\mathbf{X}_p - \sqrt{\beta_p}\mathbf{S}_p\mathbf{H}\right)}.$$

$$\tag{5.33}$$

For a realization of $\left(a, \tilde{\mathbf{f}}, \tilde{\mathbf{g}}\right)$, if $\|\mathbf{g}\|_F$ is further known, $\mathbf{R}_{\text{vec}(\mathbf{W}_p)}$ is a known matrix. From (5.33), the PDF of $\mathbf{X}_p|a, \tilde{\mathbf{f}}, \tilde{\mathbf{g}}, a_{\mathbf{g}}$ is

$$p(\mathbf{X}_p|a, \tilde{\mathbf{f}}, \tilde{\mathbf{g}}, a_{\mathbf{g}}) = \frac{1}{(2\pi)^{T_pN} \det \mathbf{R}_{\text{vec}(\mathbf{W}_p)}} e^{-\text{vec}(\mathbf{X}_p - \sqrt{\beta}\mathbf{S}_p\mathbf{H})^* \mathbf{R}_{\text{vec}(\mathbf{W}_p)}^{-1} \text{vec}(\mathbf{X}_p - \sqrt{\beta}\mathbf{S}_p\mathbf{H})}.$$

Thus,

$$\left(\hat{a}, \hat{\tilde{\mathbf{f}}}, \hat{\tilde{\mathbf{g}}}\right) = \arg\max_{(a,\tilde{\mathbf{f}},\tilde{\mathbf{g}})} p(\mathbf{X}_p|a, \tilde{\mathbf{f}}, \tilde{\mathbf{g}}, a_{\mathbf{g}})$$

$$= \arg\min_{(a,\tilde{\mathbf{f}},\tilde{\mathbf{g}})} \operatorname{vec}\left(\mathbf{X}_p - a\sqrt{\beta}\mathbf{S}_p\tilde{\mathbf{f}}\tilde{\mathbf{g}}\right)^* \mathbf{R}_{\operatorname{vec}(\mathbf{W}_p)}^{-1} \operatorname{vec}\left(\mathbf{X}_p - a\sqrt{\beta}\mathbf{S}_p\tilde{\mathbf{f}}\tilde{\mathbf{g}}\right). \quad (5.34)$$

Define

$$\eta \triangleq \frac{\alpha_p\|\mathbf{g}\|_F^2}{1+\alpha_p\|\mathbf{g}\|_F^2}. \quad (5.35)$$

After some involved calculations, the details of which can be found in [5], it can be shown that

$$\operatorname{vec}\left(\mathbf{X}_p - a\sqrt{\beta_p}\mathbf{S}_p\tilde{\mathbf{f}}\tilde{\mathbf{g}}\right)^* \mathbf{R}_{\operatorname{vec}(\mathbf{W}_p)}^{-1} \operatorname{vec}\left(\mathbf{X}_p - a\sqrt{\beta_p}\mathbf{S}_p\tilde{\mathbf{f}}\tilde{\mathbf{g}}\right)$$

$$= (1-\eta)\,\beta_p\|\mathbf{S}_p\tilde{\mathbf{f}}\|_F^2 \left(a - \frac{\Re\left(\tilde{\mathbf{f}}^*\mathbf{S}_p^*\mathbf{X}_p\tilde{\mathbf{g}}^*\right)}{\sqrt{\beta_p}\|\mathbf{S}_p\tilde{\mathbf{f}}\|_F^2}\right)^2 - (1-\eta)\frac{\Re^2\left(\tilde{\mathbf{f}}^*\mathbf{S}_p^*\mathbf{X}_p\tilde{\mathbf{g}}^*\right)}{\|\mathbf{S}_p\tilde{\mathbf{f}}\|_F^2}$$

$$+ \ \|\mathbf{X}_p\|_F^2 - \eta\|\mathbf{X}_p\tilde{\mathbf{g}}^*\|_F^2 \triangleq F\left(a,\tilde{\mathbf{f}},\tilde{\mathbf{g}}\right). \quad (5.36)$$

Thus, to minimize (5.36), we need

$$a = \frac{\Re\left(\tilde{\mathbf{f}}^*\mathbf{S}_p^*\mathbf{X}_p\tilde{\mathbf{g}}^*\right)}{\sqrt{\beta_p}\|\mathbf{S}_p\tilde{\mathbf{f}}\|_F^2}. \quad (5.37)$$

With this choice of a, from (5.36), the optimization problem in (5.34) reduces to

$$\left(\hat{\tilde{\mathbf{f}}},\hat{\tilde{\mathbf{g}}}\right) = \arg\max_{(\tilde{\mathbf{f}},\tilde{\mathbf{g}})} G\left(\tilde{\mathbf{f}},\tilde{\mathbf{g}}\right),$$

where

$$G\left(\tilde{\mathbf{f}},\tilde{\mathbf{g}}\right) \triangleq (1-\eta)\frac{\Re^2\left[\left(\mathbf{S}_p\tilde{\mathbf{f}}\right)^*\mathbf{X}_p\tilde{\mathbf{g}}^*\right]}{\|\mathbf{S}_p\tilde{\mathbf{f}}\|_F^2} + \eta\|\mathbf{X}_p\tilde{\mathbf{g}}^*\|_F^2.$$

Let

$$\mathbf{P} \triangleq \mathbf{S}_p(\mathbf{S}_p^*\mathbf{S}_p)^{-1}\mathbf{S}_p^*, \quad \mathbf{Z} \triangleq [\mathbf{P}+\sqrt{\eta}(\mathbf{I}_{T_p}-\mathbf{P})]\mathbf{X}_p. \quad (5.38)$$

We have $\left(\mathbf{S}_p\tilde{\mathbf{f}}\right)^*\mathbf{X}_p\tilde{\mathbf{g}}^* = \left(\mathbf{S}_p\tilde{\mathbf{f}}\right)^*\mathbf{P}\mathbf{X}_p\tilde{\mathbf{g}}^*$. Thus

$$G\left(\tilde{\mathbf{f}},\tilde{\mathbf{g}}\right) = (1-\eta)\frac{\Re^2\left[\left(\mathbf{S}_p\tilde{\mathbf{f}}\right)^*\mathbf{P}\mathbf{X}_p\tilde{\mathbf{g}}^*\right]}{\|\mathbf{S}_p\tilde{\mathbf{f}}\|_F^2} + \eta\|\mathbf{X}_p\tilde{\mathbf{g}}^*\|_F^2$$

$$\leq (1-\eta)\frac{\|\mathbf{S}_p\tilde{\mathbf{f}}\|_F^2\|\mathbf{P}\mathbf{X}_p\tilde{\mathbf{g}}^*\|_F^2}{\|\mathbf{S}_p\tilde{\mathbf{f}}\|_F^2} + \eta\|\mathbf{X}_p\tilde{\mathbf{g}}^*\|_F^2$$

$$= (1 - \eta)\|\mathbf{PX}_p\tilde{\mathbf{g}}^*\|_F^2 + \eta\|\mathbf{X}_p\tilde{\mathbf{g}}^*\|_F^2 \triangleq H(\tilde{\mathbf{g}}). \qquad (5.39)$$

For the second step in (5.39), Cauchy-Schwarz inequality is used. The equality in (5.39) holds when $\mathbf{S}_p\tilde{\mathbf{f}} = \gamma\mathbf{PX}_p\tilde{\mathbf{g}}^*$ for some real γ. So, when $\tilde{\mathbf{f}} = \gamma(\mathbf{S}_p^*\mathbf{S}_p)^{-1}\mathbf{S}_p^*\mathbf{X}_p\tilde{\mathbf{g}}^*$, $G\left(\tilde{\mathbf{f}}, \tilde{\mathbf{g}}\right)$ is maximized. Since $\tilde{\mathbf{f}}$ has unit-norm, we have

$$\tilde{\mathbf{f}} = \frac{(\mathbf{S}_p^*\mathbf{S}_p)^{-1}\mathbf{S}_p^*\mathbf{X}_p\tilde{\mathbf{g}}^*}{\|(\mathbf{S}_p^*\mathbf{S}_p)^{-1}\mathbf{S}_p^*\mathbf{X}_p\tilde{\mathbf{g}}^*\|_F}. \qquad (5.40)$$

With this choice of $\tilde{\mathbf{f}}$, the problem becomes the maximization of $H(\tilde{\mathbf{g}})$ defined in (5.39). Notice that $\mathbf{P}$ is a projection matrix. Thus $(\mathbf{PX}_p)^*[(\mathbf{I}_{T_p} - \mathbf{P})\mathbf{X}_p] = \mathbf{0}$. Then

$$\|\mathbf{X}_p\tilde{\mathbf{g}}^*\|_F^2 = \|\mathbf{PX}_p\tilde{\mathbf{g}}^* + (\mathbf{I}_{T_p} - \mathbf{P})\mathbf{X}_p\tilde{\mathbf{g}}^*\|_F^2 = \|\mathbf{PX}_p\tilde{\mathbf{g}}^*\|_F^2 + \|(\mathbf{I}_{T_p} - \mathbf{P})\mathbf{X}_p\tilde{\mathbf{g}}^*\|_F^2.$$

We have

$$\begin{aligned} H(\tilde{\mathbf{g}}) &= \|\mathbf{PX}_p\tilde{\mathbf{g}}^*\|_F^2 + \eta\left(\|\mathbf{X}_p\tilde{\mathbf{g}}^*\|_F^2 - \|\mathbf{PX}_p\tilde{\mathbf{g}}^*\|_F^2\right) \\ &= \|\mathbf{PX}_p\tilde{\mathbf{g}}^*\|_F^2 + \eta\|(\mathbf{I}_{T_p} - \mathbf{P})\mathbf{X}_p\tilde{\mathbf{g}}^*\|_F^2 \\ &= \|\left[\mathbf{P} + \sqrt{\eta}(\mathbf{I}_{T_p} - \mathbf{P})\right]\mathbf{X}_p\tilde{\mathbf{g}}^*\|_F^2 = \|\mathbf{Z}\tilde{\mathbf{g}}^*\|_F^2, \end{aligned} \qquad (5.41)$$

where in the last step, the definition of $\mathbf{Z}$ in (5.38) is used. It is clear that $H(\tilde{\mathbf{g}})$ is maximized when $\tilde{\mathbf{g}}^*$ is the right singular vector of the largest singular value of $\mathbf{Z}$. With this optimal choice of $\tilde{\mathbf{g}}$, by using this results in (5.40) then (5.37), the ML estimations of $\tilde{\mathbf{f}}$ and a can be obtained. The results are summarized in the following paragraph.

ML estimation with known $\|\mathbf{g}\|_F$: Let $\sigma_{\mathbf{Z},\max}$ be the largest singular value of $\mathbf{Z}$ and $\mathbf{v}_{\mathbf{Z},\max}$ be the right singular vector corresponding to the singular value $\sigma_{\mathbf{Z},\max}$. The ML estimations of $a, \tilde{\mathbf{f}}, \tilde{\mathbf{g}}$ are

$$\hat{a} = \frac{1}{\sqrt{\beta_p}}\|(\mathbf{S}_p^*\mathbf{S}_p)^{-1}\mathbf{S}_p^*\mathbf{X}_p\mathbf{v}_{\mathbf{Z},\max}\|_F, \qquad (5.42)$$

$$\hat{\tilde{\mathbf{f}}} = \frac{(\mathbf{S}_p^*\mathbf{S}_p)^{-1}\mathbf{S}_p^*\mathbf{X}_p\mathbf{v}_{\mathbf{Z},\max}}{\|(\mathbf{S}_p^*\mathbf{S}_p)^{-1}\mathbf{S}_p^*\mathbf{X}_p\mathbf{v}_{\mathbf{Z},\max}\|_F}, \quad \hat{\tilde{\mathbf{g}}} = \mathbf{v}_{\mathbf{Z},\max}^*. \qquad (5.43)$$

Thus, the ML estimation of the end-to-end channel matrix $\mathbf{H}$ is

$$\hat{\mathbf{H}} = \hat{a}\hat{\tilde{\mathbf{f}}}\hat{\tilde{\mathbf{g}}} = \frac{1}{\sqrt{\beta_p}}(\mathbf{S}_p^*\mathbf{S}_p)^{-1}\mathbf{S}_p^*\mathbf{X}_p\mathbf{v}_{\mathbf{Z},\max}\mathbf{v}_{\mathbf{Z},\max}^*. \qquad (5.44)$$

The estimation in (5.44) is based on the ideal assumption that $\|\mathbf{g}\|_F$ is known. In reality, $\|\mathbf{g}\|_F$ is unknown. Thus η is unknown and the estimations in (5.44) cannot be calculated. To obtain an estimation, we replace $\|\mathbf{g}\|_F^2$ with its mean, i.e., $\|\mathbf{g}\|_F^2 \approx \mathbb{E}(\|\mathbf{g}\|_F^2) = N$. With this approximation,

$$\eta \approx \frac{N\alpha_p}{1 + N\alpha_p}. \tag{5.45}$$

An estimation of $\mathbf{H}$ can thus be found using (5.38) and (5.44). Generally speaking, this estimation is not the exact ML estimation but an approximate one. In the following, several remarks on the SVD-based approximate ML estimation are provided.

1. The estimation $\hat{\mathbf{H}}$ in (5.44) is rank-1, which agrees with the structure of $\mathbf{H}$ in (5.26).
2. For the proposed estimation to be valid, $\mathbf{S}_p^*\mathbf{S}_p$ must be invertible. This implies a condition on the training time: $T_p \geq M$. The same requirement on the training time was derived for the ML channel estimation in a multiple-antenna system [2] and the entry-based ML estimation in (5.29).
3. The SVD-based estimation uses the rank-1 structure of the end-to-end channel matrix. For networks with single transmit and single receive antenna, the end-to-end channel is 1-dimensional, which is always rank-1. Thus the SVD-based estimation is more favorable than entry-based estimations for networks with multiple transmit and/or multiple receive antennas.
4. For the special case that $T_p = M$ and $\mathbf{S}_p$ is nonsingular, $\mathbf{P} = \mathbf{I}_{T_p}$ and $\mathbf{Z} = \mathbf{X}_p$ regardless of the value of $\|\mathbf{g}\|_F$. In this case, the estimation in (5.44) is independent in $\|\mathbf{g}\|_F$. As a result, the approximate ML estimation becomes the exact ML estimation. This is represented in the following theorem.

Theorem 5.4. *When $T_p = M$ and the pilot matrix $\mathbf{S}_P$ is nonsingular, let $\sigma_{\mathbf{X}_p,\max}$ be the largest singular value of $\mathbf{X}_p$ and $\mathbf{u}_{\mathbf{X}_p,\max}$ and $\mathbf{v}_{\mathbf{X}_p,\max}$ be the left and right singular vectors corresponding to the singular value $\sigma_{\mathbf{X}_p,\max}$. The ML estimation of $(a, \tilde{\mathbf{f}}, \tilde{\mathbf{g}})$ is*

$$\hat{a} = \frac{\sigma_{\mathbf{X}_p,\max}}{\sqrt{\beta_p}}\|\mathbf{S}_p^{-1}\mathbf{u}_{\mathbf{X}_p,\max}\|_F, \quad \hat{\tilde{\mathbf{f}}} = \frac{\mathbf{S}_p^{-1}\mathbf{u}_{\mathbf{X}_p,\max}}{\|\mathbf{S}_p^{-1}\mathbf{u}_{\mathbf{X}_p,\max}\|_F}, \hat{\tilde{\mathbf{g}}} = \mathbf{v}_{\mathbf{X}_p,\max}^*. \tag{5.46}$$

The ML estimation of the end-to-end channel matrix is thus

$$\hat{\mathbf{H}} = \frac{\sigma_{\mathbf{X}_p,\max}}{\sqrt{\beta_p}}\mathbf{S}_p^{-1}\mathbf{u}_{\mathbf{X}_p,\max}\mathbf{v}_{\mathbf{X}_p,\max}^*. \tag{5.47}$$

It can be seen from Theorem 5.4 that when $T_p = M$ and the pilot matrix is nonsingular, the ML estimation on a is a scaled version of the largest singular value of $\mathbf{X}_p$, the ML estimation on the direction of $\mathbf{g}$ is the Hermitian of the corresponding right singular vector, and the ML estimation on the direction of $\mathbf{f}$ is the corresponding left singular vector transformed by the inverse of the pilot.

Simulation on MSE

Now, we show the simulated MSE on $\mathbf{H}$, denoted as MSE($\mathbf{H}$), for the entry-based ML estimation in (5.29), the entry-based LMMSE estimation in (5.30), and the SVD-based ML estimation in (5.47). In all simulations, the channels and noises are generated independently following $\mathcal{CN}(0, 1)$. We also set $T_p = M$ and $P_s = P_r$.

First, networks with single relay antenna and different transmit and receive antennas are considered. For the training pilot, we set $\mathbf{S}_p = \mathbf{I}_M$. The following observations can be obtained from Fig. 5.7.

1. At low training power, entry-based LMMSE estimation achieves lower MSE than entry-based ML estimation. At high transmit power, the two have similar MSE performance.
2. For all simulated training power values, the SVD-based ML estimation has lower MSEs than the entry-based ML estimation. The difference is about 0.5, 1.2, and 2 dB for networks with 2, 4, and 6 transmit and receive antennas.
3. The SVD-based ML estimation is worse than the entry-based LMMSE estimation for low training power (since LMMSE estimation is better than ML estimation in the low SNR regime) but better for high training power.
4. All MSEs decreases linearly in the training power, i.e., have the scaling $\mathcal{O}(1/P)$.

This figure shows that SVD-based ML estimation is superior to entry-based estimation for multiple-antenna relay network.

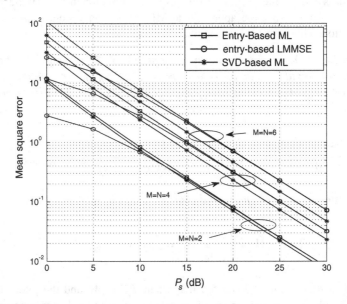

Fig. 5.7 MSEs of end-to-end channel estimations for networks with single relay antenna while $M = N = 2$, $M = N = 4$, $M = N = 6$

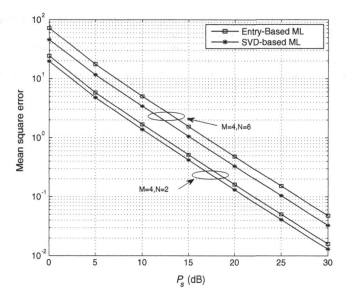

Fig. 5.8 MSEs of end-to-end channel estimations for networks with single relay antenna while $M = 4$, $N = 2$ and $M = 4$, $N = 6$

The next simulation is on networks with $T_p = M = 4$, $N = 2$ and $T_p = M = 4$, $N = 6$. Figure 5.8 shows the MSEs of entry-based ML estimation and SVD-based ML estimation. Similar phenomenon can be obtained. The SVD-based estimation is superior for all simulated training power.

References

1. Gao F, Cui T, and Nallanathan A (2008) On channel estimation and optimal training design for amplify and forward relay networks. IEEE T Wireless Communications, 7:1907–1916.
2. Hassibi B and Hochwald BM (2003) How much training is needed in multiple-antenna wireless links? IEEE T on Information Theory, 49:951–963.
3. Horn RA and Johnson CR (1986) Matrix Analysis. Cambridge University Press.
4. Jafarkhani H (2005) Space-Time Coding: Theory and Practice. Cambridge Academic Press.
5. Jing Y and Yu X (2012) ML channel estimations for non-regenerative MIMO relay networks. IEEE J Selected Areas in Communications, 30: 1428–1439.
6. Kay SM (1993) Fundamentals of Statistical Signal Processing, Volume I: Estimation Theory, Prentice Hall.
7. Nabar RU, Bölcskei HB, Kneubuhler FW (2004) Fading relay channels: Performance limits and space-time signal design. IEEE Journal on Selected Areas in Communications, 22:1099–1109.
8. Patel C, Stüber G, and Pratt T (2006) Statistical properties of amplify and forward relay fading channels, IEEE T on Vehicular Technology, 55:1–9.
9. Sun S (2011) Channel Training and Decodings for MIMO Relay Networks. M.Sc. thesis, University of Alberta.

10. Sun S and Jing Y (2011) Channel training design in amplify-and-forward MIMO relay net-
    works. IEEE T Wireless Communications, 10:3380–3391.
11. Sun S and Jing Y (2012) Training and decoding for cooperative network with multiple relays
    and receive antennas. IEEE T Communications, 50:1534–1544.
12. Tarokh V, Jafarkhani H, and Calderbank AR (1999) Space-time block codes from orthogonal
    designs. IEEE T on Information Theory, 45:1456–1467.